Assessing the Association Between Airmen Participation in Force Support Squadron Programs and Unit Cohesion

An Evaluation of the UNITE Initiative

STEPHANIE BROOKS HOLLIDAY, SARAH O. MEADOWS,
STEPHANI L. WRABEL, LAURA WERBER, CHRISTOPHER JOSEPH DOSS,
WING YI CHAN, LU DONG, BRANDON CROSBY

Prepared for the Department of the Air Force
Approved for public release; distribution unlimited

For more information on this publication, visit **www.rand.org/t/RRA554-1**.

About RAND

The RAND Corporation is a research organization that develops solutions to public policy challenges to help make communities throughout the world safer and more secure, healthier and more prosperous. RAND is nonprofit, nonpartisan, and committed to the public interest. To learn more about RAND, visit www.rand.org.

Research Integrity

Our mission to help improve policy and decisionmaking through research and analysis is enabled through our core values of quality and objectivity and our unwavering commitment to the highest level of integrity and ethical behavior. To help ensure our research and analysis are rigorous, objective, and nonpartisan, we subject our research publications to a robust and exacting quality-assurance process; avoid both the appearance and reality of financial and other conflicts of interest through staff training, project screening, and a policy of mandatory disclosure; and pursue transparency in our research engagements through our commitment to the open publication of our research findings and recommendations, disclosure of the source of funding of published research, and policies to ensure intellectual independence. For more information, visit www.rand.org/about/research-integrity.

RAND's publications do not necessarily reflect the opinions of its research clients and sponsors.

Published by the RAND Corporation, Santa Monica, Calif.

Library of Congress Cataloging-in-Publication Data is available for this publication.

ISBN: 978-1-9774-0714-6

Cover: U.S. Air Force photo/Senior Airman Ryan Lackey.

About This Report

In 2016, Air Force Chief of Staff Gen David L. Goldfein released a letter describing squadrons as "the beating heart of the United States Air Force; our most essential team" and kicked off an effort to revitalize squadrons, promoting their readiness and resilience. As part of this effort, the Air Force Services Center established the UNITE initiative, designed to provide squadron and unit commanders with opportunities to leverage Force Support Squadron (FSS) activities to increase unit cohesion. Improved unit cohesion is expected to improve readiness and resilience among unit members.

The Air Force asked RAND Project AIR FORCE to develop an evaluation that would assess the relationship between participation in FSS activities and unit and airman readiness and resilience. The UNITE Initiative provided the context for this evaluation. In this report, we provide a brief overview of the construct of unit[1] (or group) cohesion, describe the UNITE Initiative, and develop a logic model and evaluation framework. The evaluation presented in this report uses both quantitative and qualitative data from numerous sources to assess the implementation of UNITE and its impact on the intended outcomes of the initiative. After presenting the results of the evaluation, we provide the Air Force with policy implications and recommendations designed to improve the efficiency and effectiveness of the UNITE Initiative. We also outline areas for future exploration.

It is important to keep in mind that the UNITE Initiative described in this report was implemented prior to the coronavirus disease 2019 (COVID-19) pandemic in the United States.

Familiarity with the 2019 RAND report *Air Force Morale, Welfare, and Recreation Programs and Services: Contribution to Airman and Family Resilience and Readiness* (Meadows et al., 2019) is recommended (but not required) background for the current study.

The research reported here was commissioned by Air Force Manpower, Personnel and Services, Directorate of Services (AF/A1S) and conducted within the Workforce, Development, and Health Program of RAND Project AIR FORCE as part of a fiscal year 2019 project "Correlation Between Participation in FSS Programs, Services, and Activities and Airman and Unit Readiness."

RAND Project AIR FORCE

RAND Project AIR FORCE (PAF), a division of the RAND Corporation, is the Department of the Air Force's (DAF's) federally funded research and development center for studies and

[1] Although the Air Force does not have an official definition of a *unit*, we use the term generically throughout the report to represent several types of groups of airmen, including groups, squadrons, flights, and elements.

analyses, supporting both the United States Air Force and the United States Space Force. PAF provides DAF with independent analyses of policy alternatives affecting the development, employment, combat readiness, and support of current and future air, space, and cyber forces. Research is conducted in four programs: Strategy and Doctrine; Force Modernization and Employment; Resource Management; and Workforce, Development, and Health. The research reported here was prepared under contract FA7014-16-D-1000.

Additional information about PAF is available on our website: www.rand.org/paf/

This report documents work originally shared with DAF on July 14, 2020. The draft report, issued on June 25, 2020, was reviewed by formal peer reviewers and DAF subject-matter experts.

Contents

Figures

Tables

Summary

Issue

The UNITE Initiative provides unit leadership with funding to participate in cohesion-promoting activities, often by using activities provided by the installation's Force Support Squadron (FSS). The Air Force asked RAND to evaluate whether participation in UNITE events is associated with increased unit cohesion, which could ultimately impact airmen and unit readiness and resilience.[2]

Approach

We developed a logic model for UNITE to guide the evaluation (Figure S.1), which visually describes how a program operates to achieve an expected outcome. We conducted a process evaluation to understand program implementation and identify successes, limitations, and lessons learned. An outcome evaluation explored whether UNITE participation was associated with expected short-term and intermediate outcomes. Data sources included interviews and open-ended surveys with community cohesion coordinators (C3s), who actually implement UNITE; post-participation surveys completed by airmen; and event details submitted by unit commanders.

Key Findings

- C3s were seen as valuable resources for units, but they experienced some practical barriers to completing their jobs, such as changing policies, unwieldy data submission processes, and limited marketing materials. Aa lack of awareness on the part of commanders may also have hindered uptake of the initiative.
- Airmen and commanders who participated in UNITE were largely satisfied with UNITE. In their view, UNITE events provided an opportunity to relax and socialize among members of a unit.
- Certain types of events were more likely to be associated with expected short-term outcomes (roughly two weeks after participation). Specifically, UNITE events that were provided by the local FSS were rated as less of an opportunity to relax and decompress, engage with other airmen, and promote and reinforce peer, squadron, and Air Force values.
- UNITE events that airmen rated as an opportunity to decompress, interact with other airmen, and reinforce peer, squadron, and Air Force values were associated with higher

[2] It is important to keep in mind that the UNITE Initiative described in this report was implemented prior to the coronavirus disease 2019 (COVID-19) pandemic in the United States.

ratings of overall unit cohesion two weeks after participation. However, the association between social interaction and overall cohesion faded over time and was no longer significant by six weeks after the event.

Recommendations

- The Air Force could increase uptake of UNITE through centralized marketing tools and increased marketing resources for C3s. Materials could emphasize that UNITE is intended to enhance unit cohesion. In addition, clearer and more consistent guidance on UNITE policies and procedures would benefit C3s, commanders, and airmen.
- Data systems used to collect UNITE event data are not user friendly, limiting the data available for both programmatic and evaluation purposes. Data quality could be improved by designing interfaces for report submission that are more user friendly or using tools (such as Common Access Card readers) at events to track participation. Better data might also shed light on the negative association between FSS-provided activities and certain short-term outcomes.
- Sustaining cohesion-related improvements that are associated with UNITE events will likely require repeated events that provide airmen with the opportunity to unwind, interact with peers, and promote unit, squadron, and Air Force values.

The next step in the evaluation, should the Air Force decide to continue the initiative, is to link UNITE events to the impacts noted in the logic model: airmen and unit readiness and resilience. Such an analysis will require more-consistent and more-nuanced data about events and participants.

Figure S.1. UNITE Logic Model

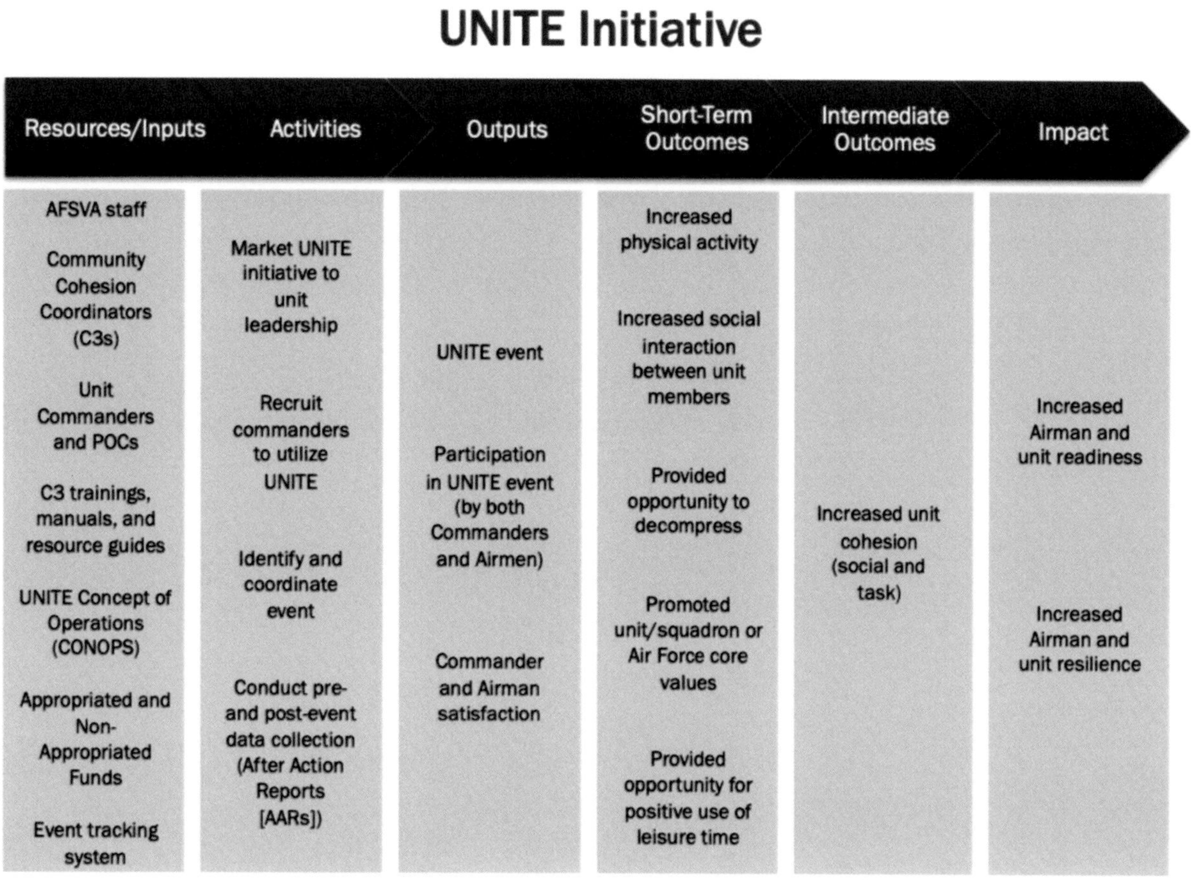

UNITE Initiative

Resources/Inputs	Activities	Outputs	Short-Term Outcomes	Intermediate Outcomes	Impact
AFSVA staff Community Cohesion Coordinators (C3s) Unit Commanders and POCs C3 trainings, manuals, and resource guides UNITE Concept of Operations (CONOPS) Appropriated and Non-Appropriated Funds Event tracking system	Market UNITE initiative to unit leadership Recruit commanders to utilize UNITE Identify and coordinate event Conduct pre- and post-event data collection (After Action Reports [AARs])	UNITE event Participation in UNITE event (by both Commanders and Airmen) Commander and Airman satisfaction	Increased physical activity Increased social interaction between unit members Provided opportunity to decompress Promoted unit/squadron or Air Force core values Provided opportunity for positive use of leisure time	Increased unit cohesion (social and task)	Increased Airman and unit readiness Increased Airman and unit resilience

NOTE: AFSVA = Air Force Services Agency.

Acknowledgments

The research team would like to thank Michael Coltrin, our study point of contact, as well as personnel at the Air Force Services Center, Air Force Personnel Center, and Air Force Survey Office. We extend our appreciation to the C3s who contributed to the study by participating in our interviews and survey. We are also grateful to Ray Conley and Kirsten Keller for their helpful comments on earlier versions of this report and for management support. We would like to thank Daniel Kim, Laura King, Dic Donohue, Tina Petrossian, and Clare McCaffrey for their assistance in the preparation of this report. We would also like to thank our reviewers, Thomas Trail and Terri Tanielian.

Abbreviations

AAR	after action report
AF/A1S	Air Force Manpower, Personnel and Services, Directorate of Services
AFI	Air Force Instruction
AFPC	Air Force Personnel Center
AFSVC	Air Force Services Center
APF	appropriated funds
C3	Community Cohesion Coordinator
CONOPS	concept of operations
CONUS	continental United States
CSBT	collaborative story building and telling
DCS	Defense Collaboration Service
DoD	U.S. Department of Defense
DoD ID	U.S. Department of Defense identification number
FSS	Force Support Squadron
GEQ	Group Environment Questionnaire
ITT	Information, Tickets, and Travel
MAJCOM	major command
MOA	Memorandum of Agreement
MWR	Morale, Welfare, and Recreation
NAF	nonappropriated funds
OCONUS	outside the continental United States
PDMS	personal disclosure mutual sharing
POC	point of contact
QED	quasi-experimental design
RCT	randomized control trial
RTE	ready-to-execute activity

Chapter 1. Introduction

In 2016, Air Force Chief of Staff Gen David L. Goldfein released a letter describing squadrons as "the beating heart of the United States Air Force; our most essential team" (Goldfein, 2016). In this letter, he initiated an effort to revitalize squadrons, which had an implementation plan with three areas of focus: (1) the mission, (2) strengthening leadership and culture, and (3) taking care of airmen and families (Barnett, 2018). Ultimately, the squadron revitalization effort aimed to promote the readiness and resilience of the force (Brissett, 2017).

Within the Air Force, Force Support Squadrons (FSSs) have the primary mission of assisting units and individuals in building on the tenets of Comprehensive Airman Fitness by providing programs, services, and experiences that foster ready and resilient airmen. In light of the squadron revitalization effort, the Air Force Services Center (AFSVC) established the UNITE Initiative, which is designed to provide squadron and unit commanders with opportunities to leverage FSS activities to increase unit cohesion. Improved unit cohesion is expected to improve readiness and resilience among unit members. The Air Force asked RAND Project Air Force researchers to develop and execute an evaluation that would assess the relationship between participation in FSS activities and unit and airman readiness and resilience. The UNITE Initiative provided the context for this evaluation. In this chapter, we provide a brief overview of the construct of unit[3] (or group) cohesion, describe the UNITE Initiative, and present an overview of the study.

It is important to keep in mind that the UNITE initiative described in this report was implemented prior to the coronavirus disease 2019 (COVID-19) pandemic in the United States.

Unit (or Group) Cohesion

Unit cohesion can be considered a category of group cohesion that is specific to the military. Group cohesion represents the "tendency for a group to stick together and remain united in the pursuit of its instrumental objectives and/or for the satisfaction of member affective needs" (Carron, Brawley, and Widmeyer, 1998, p. 213, as cited in Carron and Brawley, 2000). Group cohesion has multiple dimensions, including the following:[4]

- *Task cohesion* is the shared commitment among members to achieving a goal that requires the collective efforts of the group. A group with high

---

[3] Although the Air Force does not have an official definition of a *unit*, we use the term generically throughout the report to represent several types of groups of airmen, including groups, squadrons, flights, and elements.

[4] *Group pride*, a dimension discussed less frequently in the literature, is the "extent to which group members exhibit liking for the status or the ideologies that the group supports or represents, or the shared importance of being a member of this group" (Beal et al., 2003, p. 995). We do not further explore group pride in this report.

task cohesion is composed of members who share a common goal and who are motivated to coordinate their efforts as a team to achieve that goal.

- *Social cohesion* is the extent to which group members like each other, prefer to spend their social time together, enjoy each other's company, and feel emotionally close to one another (National Defense Research Institute, 2010, p. 139).

The group cohesion literature generally focuses on the relationship between group cohesion and group performance, offering the hypothesis that groups are more effective when members perceive their group to be cohesive (e.g., Evans and Dion, 2012). A meta-analysis of group cohesion in military units found that cohesion was positively associated with group and individual performance, as well as individual well-being, retention, and readiness (Oliver et al., 1999). Beal et al., 2003, notes that although all dimensions of cohesion are associated with group performance, the relationship between task cohesion and performance is largest in magnitude.

Because of this association between cohesion and group performance, finding ways to promote unit cohesion is a priority for the military. Prior research on cohesion in the military (Rostker et al., 1993; National Defense Research Institute, 2010) has identified mechanisms through which unit cohesion develops, such as shared group membership; similarity of attitudes, interests, and values; shared experiences; and high-quality leadership and training. These are factors that have a demonstrated relationship with group cohesion in military contexts. However, there have been few studies to formally test the degree to which interventions or programs that use these principles actually serve to improve cohesion. Similarly, few studies have examined whether increases in unit cohesion result in downstream effects on resilience and readiness.

UNITE Initiative

The UNITE Initiative was established by the AFSVC as part of the effort to revitalize squadrons. Through UNITE, unit commanders have access to funding to support activities that are designed to improve unit cohesion. The UNITE Initiative was implemented in fall 2018 at an initial set of 41 installations, representing a diverse variety of geographic locations and major commands (MAJCOMs). UNITE was rolled out at the remaining installations in early 2019.

To better understand the UNITE Initiative, we held discussions with key contacts within Air Force Manpower, Personnel and Services, Directorate of Services (AF/A1S) and AFSVC and reviewed relevant materials (e.g., the UNITE Concept of Operations [CONOPS]) during the months leading up to implementation of the initiative and throughout the first year of implementation.

UNITE Initiative Staff and Training

UNITE is overseen and executed by staff at AFSVC. This includes oversight by the AFSVC Community Programs Branch Chief and two Community Program and Unit Cohesion Lead

Coordinators, who are located at AFSVC. To implement UNITE at the installation level, AFSVC hired Community Cohesion Coordinators (C3s) across participating installations. Per the UNITE CONOPS, C3s are responsible for planning programs, activities, and events that directly support unit cohesion, leveraging Morale, Welfare, and Recreation (MWR) programming along with resources and activities in the local community. C3s participate in an in-residence, weeklong centralized training program provided by AFSVC. During the training, C3s learn more about the scope of their role, gain hands-on practice by planning a sample UNITE event, receive training on the financial aspects of UNITE, and interact with other C3s. C3s are expected to develop a set of standardized event options that can be offered to units, but they can also work more closely with units to develop a tailored experience that meets a unit's specific needs.

C3s work directly with units to plan events. Units assign a point of contact (POC), who serves as the liaison between the unit commander and/or leadership and the C3. The POC is responsible for working with the C3 to develop an event idea that addresses the needs of their unit (more detail about the C3 and POC roles is presented in Chapter 3). C3s conduct outreach to commanders and POCs to explain the initiative and to encourage units to take advantage of the funding.

UNITE Events and Funding

UNITE is designed to serve active-duty, nonactivated reserve component, and civilian airmen. Per the UNITE Phasing CONOPS, the implementation of UNITE took place in phases according to the availability of funding, beginning with active-duty service members and moving to nonactivated Air Force Reserve Component service members. UNITE events must be designed with a cohesion or team-building component. As noted, C3s work with unit POCs to develop an event proposal, which is then submitted to AFSVC. If the cohesion or team-building component of the event is not clear, AFSVC requires the event to more clearly include these elements. AFSVC staff ultimately must approve each event. During the first year of implementation, events were submitted for approval to AFSVC via a UNITE SharePoint website. The event request form required information about the details of the event, reasons for participation (e.g., to increase morale, camaraderie, or esprit de corps; work on a team-building exercise; or develop a new skill or competency), estimated number of participants, and estimated funding required, including appropriated funds (APF) and nonappropriated funds (NAF). After the event, C3s are responsible for submitting an after action report (AAR), which includes information collected from POCs and commanders about actual attendance, funding used, and perceptions about whether the event was successful.

Through UNITE, squadrons have access to two types of funding for events. APF, provided via Memorandum of Agreement (MOA), are available to offset participation costs (the amounts varied during initial implementation),[5] whereas NAF are provided to cover food and beverages for UNITE events ($5 per participant). AFSVC provides C3s with suggested events that could be implemented. For example, units could make use of MWR facilities, such as outdoor recreation, golf courses, or bowling centers; participate in outdoor adventures, such as rock climbing or ropes courses; or participate in team challenges, such as combat bowling or pool obstacle courses. However, UNITE events do not necessarily require funding: For example, a UNITE event could bring together unit members to participate in a volunteer activity in the local community. The goal is for each unit to participate in at least one UNITE event per year. Commanders have discretion with respect to the size of the unit that participates in a given event; during the first year of implementation, the event supported large, higher-level units (e.g., through a squadron fun day or picnic) but also supported smaller units (e.g., through elements or workstations). Note that, throughout the report, when we refer to a *unit*, we use the term broadly to indicate any of these levels of organization.

Present Study

FSS activities are designed to promote readiness and resilience in service members and their families. In part, FSSs accomplish this by offering activities and programs that could be used to promote unit cohesion. For example, services such as outdoor recreation, golf courses, and bowling centers provide units with the opportunity to engage in group activities. FSSs also oversee programs that are specifically designed to promote resilience and readiness, such as Recharge for Resiliency, which leverages MWR activities to implement program components focused on single airmen, families, and units.

Like other FSS activities, the UNITE Initiative aims to promote resilience and readiness through a focus on unit cohesion. Previous research (such as the studies described earlier) suggests that providing units with the opportunities to participate in group activities could serve to improve cohesion. However, the Air Force lacks data that demonstrate a correlation between the use of these activities and expected outcomes. Therefore, there is a need to understand whether and how FSS experiences might foster unit cohesion, readiness, and resilience, and there is a need to collect data to explore whether there is a link between individual and unit participation in FSS programs and readiness indicators. With this study, we aimed to examine this connection by conducting an initial evaluation of the UNITE Initiative. In certain ways, this evaluation was also designed to serve as a proof of concept regarding the ability to collect data

---

[5] The amount of APF funding per participant changed over the course of the study. Installations participating in the first wave of implementation had access to $17.50 per participant for events. After the initiative expanded, an additional subset of installations had access to $13.00 per participant from August to December 2019. The official second phase of UNITE began in January 2020, at which time the number of installations receiving funding expanded and the amount decreased to $13.50 per participant across installations. Thus, in some of the quotes used later in the report, both $13.50 and $17.50 are referenced. Note that NAF funding remained unchanged.

from airmen participating in UNITE across all installations and the use of those data for evaluation purposes.

This report describes the results of our evaluation of UNITE. In Chapter Two, we present an evidence-informed framework describing the ways that UNITE could be expected to foster unit cohesion, readiness, and resilience. The results from this chapter are used to inform our evaluation of UNITE. In Chapter Three, we present the results of our evaluation of the implementation of UNITE (also known as a process evaluation), which we accomplished by conducting interviews with C3s and reviewing post-event feedback from C3s, units, and airmen. The goal of the process evaluation was to understand how the program was implemented and identify successes, limitations, and lessons learned. In Chapter Four, we present the results of an outcome evaluation, which explored whether UNITE participation was achieving its intended outcomes, using post-participation surveys completed by airmen who participated in UNITE to examine the outcomes of participation, including the perceived influence on unit cohesion. The report also contains four appendixes:

- Appendix A provides a detailed review of our literature review on cohesion and interventions designed to improve group cohesion
- Appendix B provides a review of the methods used in our qualitative analysis
- Appendix C provides details on the post-UNITE participation survey completed by airmen
- Appendix D provides a review of the methods used in our quantitative analysis.

Chapter 2. Developing an Evidence-Informed Framework for UNITE

The goal of UNITE is to promote unit cohesion, which is expected to foster ready and resilient airmen and units. To better understand the empirical foundations for an initiative like UNITE, we drew on the research literature. To begin, we conducted a review of the extant literature on unit cohesion, with the goal of (1) identifying the factors that have been demonstrated to be associated with unit cohesion and (2) identifying the types of interventions and programs that have been demonstrated to be associated with the promotion of cohesion in small groups. The detailed results of this literature review are described in Appendix A. The next step was to understand the ways that improvements in unit cohesion might lead to downstream improvements in readiness and resilience. We conducted a crosswalk of the cohesion predictors identified through the literature review with the building blocks of readiness and resilience, which were identified as part of a previous RAND study (Meadows et al., 2019). In this chapter, we provide a broad overview of the findings of the literature review and describe the results of the crosswalk process. Finally, we also present the logic model the research team developed for UNITE according to our understanding of the UNITE Initiative (see Chapter 1) and the evidence-informed framework discussed in this chapter.

Predictors of Group Cohesion

The first component of our literature search focused on identifying predictors of group cohesion (for a detailed summary of methods and findings, see Appendix A). We focused on articles that assessed cohesion as an outcome within small groups (e.g., sports teams, work units, military units). We included both qualitative and quantitative studies in our review. The majority of identified studies used cross-sectional research designs; therefore, findings from the literature review do not necessarily provide evidence of a causal relationship between the predictors and cohesion. However, the predictors identified were guided by theory.

A total of 59 articles were included in our review. From these articles, we identified the following predictors of cohesion:

- **Demographic: social characteristics.** Includes characteristics such as gender composition, age of participants, race/ethnicity, and marital status. Research has been somewhat mixed regarding the effect of group composition on cohesion (e.g., Dermatis et al., 2001; Griffith, 1989; Harrison, Price, and Bell, 1998).
- **Demographic: work/military.** Includes work-related history, such as length of membership in a group, length of employment, and diversity of work-related experience. Research has found that length of group membership is positively associated with

cohesion (e.g., Bartone et al., 2002), though other work/military demographics have more-mixed evidence (e.g., Aubke et al., 2014; Sánchez and Yurrebaso, 2009).

- **Group communication.** Includes the method of group communication (e.g., in-person versus virtual) and perceptions of communication as effective and appropriate. For example, more-effective and more-appropriate communication is associated with higher cohesion (Troth, Jordan, and Lawrence, 2012).
- **Healthy social interaction.** Includes practices such as promoting empathy and allowing for emotional expression. Some evidence suggests these practices increase cohesion (e.g., Plante, 2006).
- **Individual subjective cognitions.** Includes perceptions of the work (e.g., job satisfaction) and perceptions of the group (e.g., connection with group members, value of the group). Factors such as degree of identification with the group or its goals and perceived value of the group are associated with higher cohesion (e.g., Dermatis et al., 2001; López et al., 2015).
- **Leadership.** Includes both the characteristics and the behaviors of leaders in promoting group cohesion. Authentic, supportive, fair, and transparent leadership is associated with higher cohesion (e.g., Charbonneau and Wood, 2018; García-Guiu, Molero, and Moriano, 2015; Lee and Farh, 2004).
- **Mental health.** Includes symptoms of depression and posttraumatic stress disorder. Higher incidences of mental health symptoms among group members have been negatively associated with cohesion (Dermatis et al., 2001; Welsh, Olson, and Perkins, 2019).
- **Personality.** Includes the Big Five personality traits, attachment style, and hardiness. Evidence has been mixed for some aspects of personality, though greater emotional intelligence and hardiness are associated with greater cohesion (e.g., Bartone et al., 2002; Moore and Mamiseishvili, 2012).
- **Group culture.** Refers to the culture or climate within the group. For example, having a task-oriented team climate is associated with greater cohesion, whereas greater emphasis on individual achievement is associated with poorer cohesion (Boyd et al., 2014).
- **Group dynamics.** Refers to the interpersonal dynamics among group members. For example, greater cooperation and ability to work together toward a goal are associated with higher cohesion (e.g., Chen et al., 2009; Kaymak, 2011).
- **Shared culture.** Refers to members of a group sharing aspects of their cultural backgrounds. There is some evidence that having a shared heritage or national pride is associated with higher cohesion (e.g., Boer and Abubakar, 2014).
- **Shared experience.** There is evidence that groups that have shared experiences (such as successfully achieving goals together, participating in collaborative activities, or deploying together) have higher cohesion (e.g., Cordobés, 1997; Griffith, 1989; Kaymak, 2011).

In addition to the cross-sectional nature of these articles, there was wide variability in the methods used to measure cohesion and the populations studied. This variability limited our ability to identify which of these factors are the "strongest" predictors of cohesion or which might be the most applicable to the military context. However, this review provided an initial framework for understanding the factors that could be expected to contribute to cohesion.

Cohesion Interventions

With our literature review, we also sought to understand what types of interventions have been shown to be effective in increasing cohesion in small groups. Units that participate in UNITE have substantial discretion with respect to the nature of the events they plan. There must be a clear team-building or cohesion component to the event, but (to our knowledge) AFSVC does not use a set of objective criteria to evaluate this requirement. Through our review of the cohesion intervention literature, we were interested in identifying any activities that have been shown to be effective in promoting cohesion. We expected that this might inform our empirical framework for the evaluation but could also reveal information that would be useful to units when planning UNITE events.

Through a multipronged search strategy, we identified 29 studies of cohesion and team-building interventions (for a detailed summary of methods and findings, see Appendix A). There are key limitations to this literature. Specifically, this literature relied heavily on correlational approaches, limiting our ability to draw conclusions about the types of interventions that effectively increase cohesion in small groups. Moreover, many interventions were tested in distinct populations (e.g., university counseling students, individuals with depressive disorders), further limiting their generalizability and applicability to UNITE. These limitations preclude us from relying on the intervention literature to identify model programs or program elements.

However, a broad review of these interventions provides insight into the types of practices that other groups have used to promote cohesion.[6] Interventions included such activities as collaborative story building and telling (CSBT), a practice by which a group of people share personal experiences and then work together to construct a group narrative, or story (Treadwell et al., 2011); team goal-setting (e.g., Senécal, Loughead, and Bloom, 2008); journal-sharing, in which members of a small group write journal entries describing events in their lives (e.g., Oh et al., 2018); and leader-training (e.g., McLaren, Eys, and Murray, 2015). Some studies examined multicomponent protocols to promote team-building, which included such elements as building group norms and promoting social interactions among group members (e.g., Bruner and Spink, 2010; Carron and Spink, 1993).

Next, we examined the connections between the cohesion literature and our previous work, which was focused on the building blocks of readiness and resilience. Our goal was to develop an evidence-informed framework for understanding the ways that a program like UNITE might increase cohesion and, in turn, improve readiness and resilience. We describe this process in the next section.

---

[6] Detailed descriptions of each intervention are included in Appendix A.

Understanding the Connections Between Cohesion, Readiness, and Resilience

In a previous study that was focused on developing an evidence-informed framework for Air Force MWR programs, RAND researchers developed a model of the building blocks of resilience and readiness (Meadows et al., 2019). This model used the resilience and readiness literature and identified specific precursors of resilience and readiness (see Figure 2.1). Building blocks in the model were classified along two dimensions. First, they were categorized at the system level, building on classical ecological systems theory (Bronfenbrenner, 1979). Specifically, the model included building blocks across four system levels: individual, family, peer/squadron, and community. It also included background characteristics (e.g., sociodemographic characteristics) that might influence access to the building blocks across levels. Second, the building blocks were categorized by their proximity to resilience and readiness, including both direct and indirect building blocks. Whereas direct building blocks can be considered primary facilitators of resilience and readiness, indirect building blocks first influence a direct building block, which then contributes to resilience and readiness.

UNITE specifically targets unit cohesion as a mechanism through which readiness and resilience may be created and strengthened. However, unit cohesion was not a specific building block that emerged from the literature. Rather, aspects of unit cohesion are likely captured by several building blocks. To better understand the empirical underpinnings of the association between unit cohesion and the readiness and resilience building blocks in our model, we performed a crosswalk between the cohesion literature and the building blocks model.

This crosswalk included two stages:

- First, we identified which resilience and readiness building blocks were relevant to each cohesion predictor.[7] Not all cohesion predictors were reflected in the building blocks model, which demonstrates that some factors associated with cohesion are distinct from those that have been shown to increase resilience and readiness. In addition, some cohesion predictors had only a partial match with a particular building block (such that a cohesion predictor might not map onto all components of a specific building block), and not all aspects of a cohesion predictor were necessarily represented in the building blocks model. However, the majority of the cohesion predictors were related to at least one building block; in many cases, the predictors were related to multiple building blocks.
- Second, we integrated the findings of the cohesion predictor and cohesion intervention literature searches by identifying which cohesion predictors appear to be

[7] Three members of the research team reviewed each of the cohesion predictors and determined which building blocks were conceptually related to the predictor.

targeted by each of the interventions identified in the literature.[8] This exercise provided insight into the mechanisms by which programs and initiatives (such as UNITE) might promote cohesion, because not all cohesion predictors would be expected to change as a result of intervention. Some interventions are more targeted in scope: For example, leader training interventions are relevant to the leadership predictor. Others appear to target multiple cohesion predictors, such as team-building interventions, which serve to increase healthy social interaction and improve group dynamics.

[8] To conduct this step of the crosswalk, four members of the research team (three of whom were also part of the team who developed the building blocks model) reviewed the descriptions of each intervention and identified which predictors were targeted by the intervention.

Figure 2.1. Evidence-Informed Building Blocks Model of Resilience and Readiness

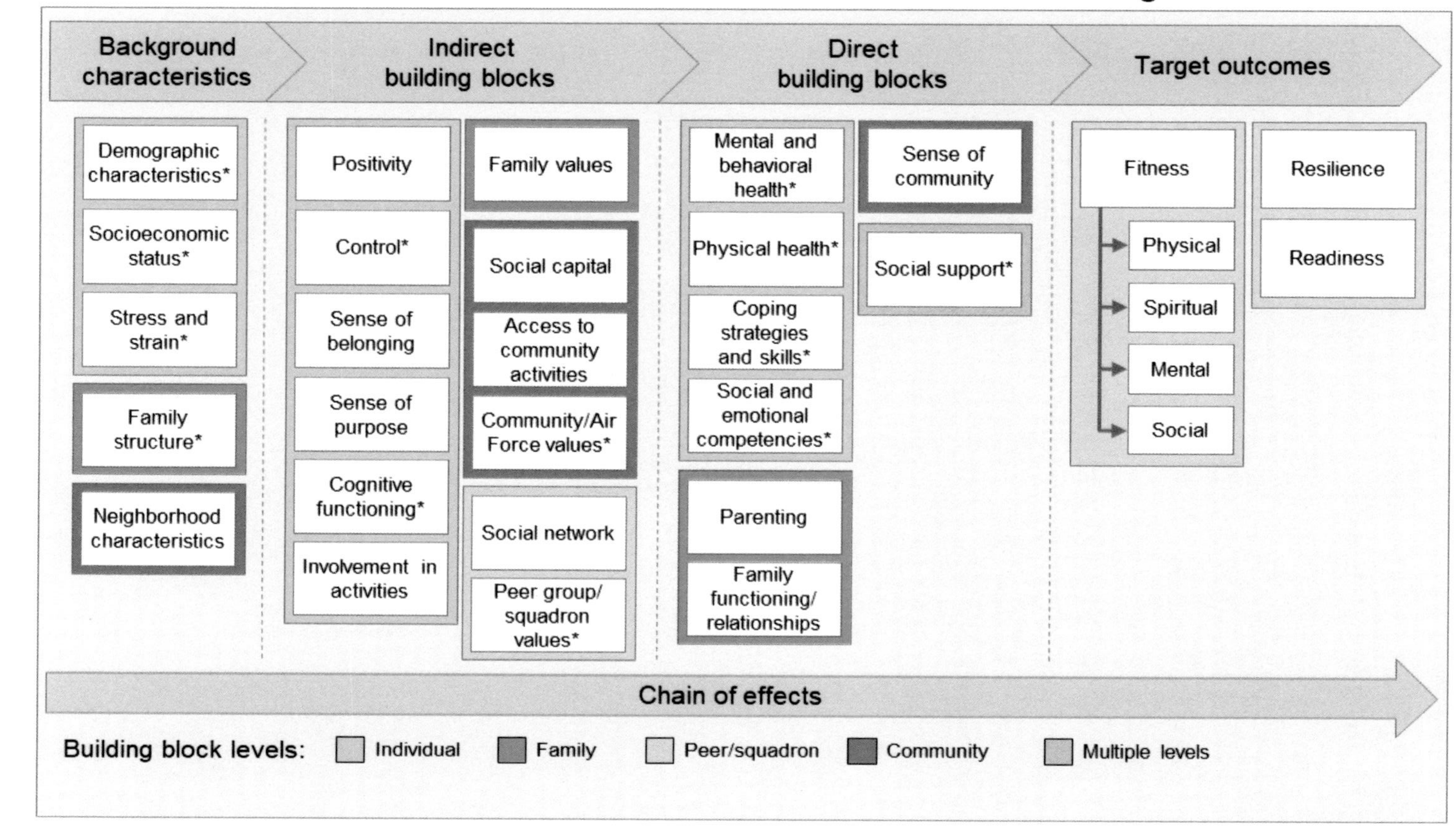

SOURCE: Meadows et al., 2019.
NOTE: * indicates the building block was identified in the readiness literature and the resilience literature.

The results of this crosswalk are presented in Table 2.1 (a version of this table that indicates the specific interventions that mapped onto each predictor appears in the appendix as Table A.3). The cohesion interventions appear to most commonly target healthy social interactions, leadership, group culture, group dynamics, and shared experiences. Individual-level characteristics (such as individual subjective cognitions, mental health, or personality) are not targeted by cohesion interventions. Instead, most interventions aimed to improve group member skills (e.g., leadership skills) or provide an opportunity for group members to work together (e.g., to interact socially, solve problems, or set goals). Interestingly, no interventions directly targeted group communication as a primary mechanism of increasing cohesion, though group communication seems like a mechanism that could easily be targeted through a cohesion program or initiative.

We found that most cohesion predictors were connected to at least one building block. We found that three cohesion predictors were matched with building blocks at the individual level, five at the peer/squadron level, six at the community level, and two at the background level. No cohesion predictors were matched with building blocks at the family level, though this is expected because our search did not target family cohesion, given that UNITE is not designed to address families (though some events include family members). The majority of the matched building blocks are at the peer/squadron and community levels.

This crosswalk helps identify two important points. First, it demonstrates the types of mechanisms that cohesion interventions target to promote cohesion. Second, it illustrates that the types of factors that predict cohesion are connected to the building blocks of resilience and readiness. This point supports the hypothesis that programs and activities that target these building blocks (including those that unit leaders select through UNITE) have the potential to increase cohesion among airmen and, ultimately, promote readiness and resilience.

Table 2.1. Crosswalk of Predictors of Cohesion, Cohesion Intervention Mechanisms, and Resilience and Readiness Building Blocks

Predictor of Cohesion	Cohesion Intervention Mechanism	Building Block
Demographic: social characteristics	--	Demographic characteristics (background)
Demographic: work/military	--	--
Group communication	--	--
Healthy social interaction	Opportunity to share personal experiences with the group	Social support (peer/squadron)
Individual subjective cognitions	--	Sense of belonging (individual) Sense of community (community)
Leadership	Training for leaders on teamwork, developing positive group norms, social interactions within group	Peer/squadron values (peer/squadron) Community/Air Force values (community)
Mental health	--	Mental and behavioral health (individual)
Personality	--	Social and emotional competencies (individual) Positivity (individual)
Group culture	Teaching team to support each other, establishing distinctiveness of the group	Sense of community (community) Social support (peer/squadron) Social network (peer/squadron) Social capital (community) Peer group/squadron values (peer/squadron) Community/Air Force values (community)
Group dynamics	Developing skills for group to work collectively to set goals, solve problems	Social network (peer/squadron) Social support (peer/squadron) Social capital (community) Sense of community (community)
Shared culture	--	Sense of community (community) Peer group/squadron values (peer/squadron) Community/Air Force values (community)
Shared experience	Providing opportunity for group to share in an activity or event	Sense of community (community) Social capital (community) Access to community activities (community) Stress and strain (background)

NOTE: Some cohesion predictors have only a partial match with the identified building blocks. A cohesion predictor might only have a direct connection with certain components of a given building block, and not all aspects of a cohesion predictor were necessarily represented in the building blocks model.

UNITE Logic Model and Evaluation Framework

A logic model visually summarizes a program's operations and outcomes (Acosta et al., 2013; Milstein, Wetterhall, and CDC Evaluation Working Group, 2000). It includes a program's *resources* or inputs (such as staff, funding, and guidance documents that govern a program); *activities* that are carried out as part of the program; *outputs*, which are the immediate results of program activities (such as participation in program activities); *short-term outcomes* (expected immediately following participation); *intermediate outcomes* (expected in the several weeks to months following participation in an event); and long-term *impacts* (longer-term effects of the program observed at the group or community level). A logic model demonstrates the hypothesized relationship between program resources and inputs, activities, and expected results or outcomes. In other words, a logic model is a visual depiction of how a program is supposed to work to achieve a desired outcome. The logic model for UNITE is presented in Figure 2.2.

The first three columns of the logic model describe the operation of the program. The resources required to implement UNITE are relevant staff, including staff located at AFSVC and installation-level C3s; unit commanders and POCs; training materials and resource guides, which are described in more detail in Chapter 3; the UNITE CONOPS; funding; and an event tracking system.

C3s have a key role in UNITE Initiative activities, which include marketing the initiative to unit leadership and recruiting units to participate, working with units to develop and execute a UNITE event, and submitting information about the event through AARs. The outputs are participation in UNITE by commanders and unit members, as well as satisfaction with the UNITE event.

The second three columns summarize the short-term outcomes expected from the program. Because UNITE events provide units with an opportunity to engage in activities focused on cohesion and team-building, the expected short-term outcomes include increased physical activity (when the event has a physical component, such as a sports tournament), increased social interaction between unit members, an opportunity to decompress, and promotion of squadron and Air Force values. UNITE events may take place during duty hours or off-duty hours; those that take place during off-duty hours are also expected to contribute to positive use of leisure time. These short-term outcomes highlight some of the predictors from our literature review (e.g., social interaction, group culture and dynamics, shared experience).

By contributing to these short-term outcomes, UNITE is expected to increase unit cohesion in the intermediate term. The ultimate intended impact of the program is to increase the resilience and readiness of individual airmen and units.

Figure 2.2. UNITE Logic Model

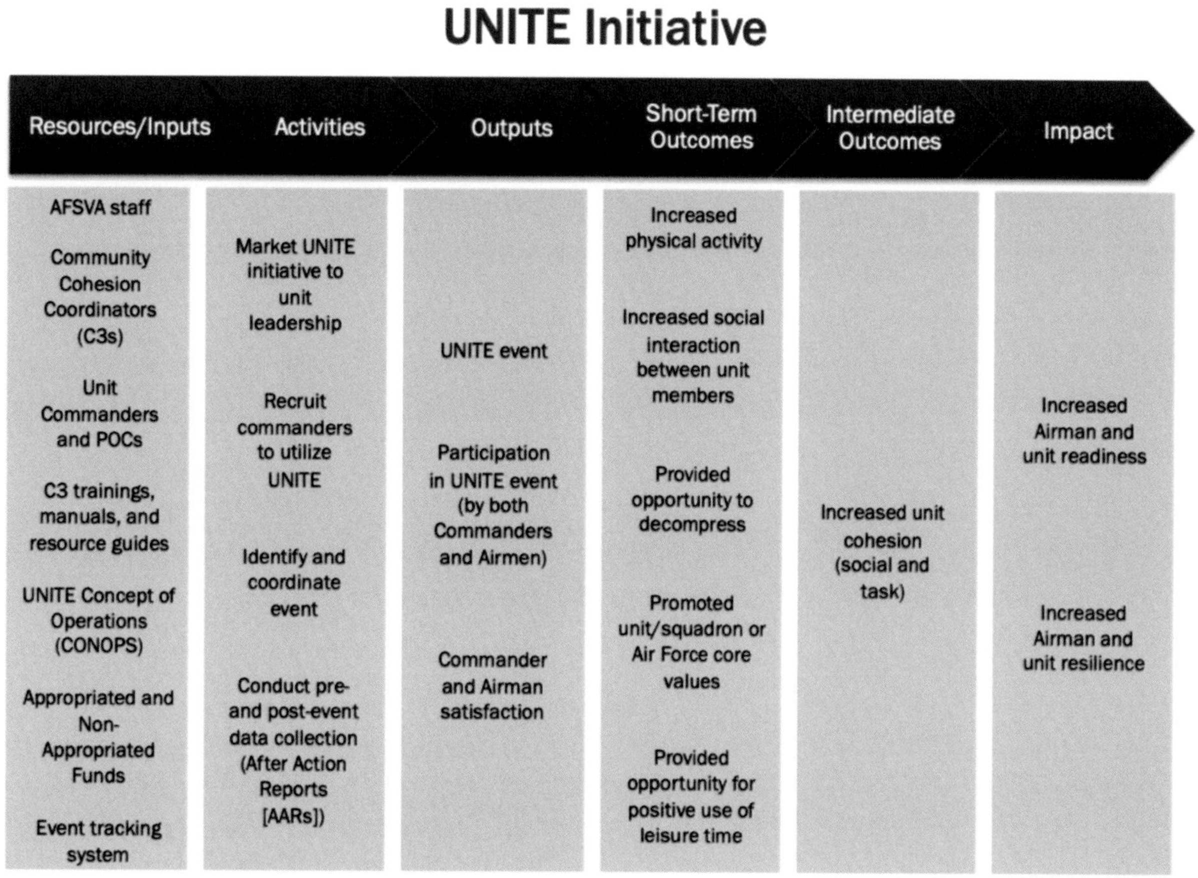

UNITE Initiative

Resources/Inputs	Activities	Outputs	Short-Term Outcomes	Intermediate Outcomes	Impact
AFSVA staff Community Cohesion Coordinators (C3s) Unit Commanders and POCs C3 trainings, manuals, and resource guides UNITE Concept of Operations (CONOPS) Appropriated and Non-Appropriated Funds Event tracking system	Market UNITE initiative to unit leadership Recruit commanders to utilize UNITE Identify and coordinate event Conduct pre- and post-event data collection (After Action Reports [AARs])	UNITE event Participation in UNITE event (by both Commanders and Airmen) Commander and Airman satisfaction	Increased physical activity Increased social interaction between unit members Provided opportunity to decompress Promoted unit/squadron or Air Force core values Provided opportunity for positive use of leisure time	Increased unit cohesion (social and task)	Increased Airman and unit readiness Increased Airman and unit resilience

NOTE: AFSVA = Air Force Services Agency.

The short-term outcomes identified in the logic model also map onto several of the readiness and resilience building blocks identified in Meadows et al., 2019. The link between the short-term outcomes of UNITE and the readiness and resilience building blocks is summarized in Table 2.2, including the conceptual rationale for the connection between each short-term outcome and the associated building blocks. This crosswalk further highlights the ways in which UNITE might ultimately contribute to readiness and resilience.

Table 2.2. Link Between UNITE Short-Term Outcomes and Readiness and Resilience Building Blocks

Intended Short-Term Outcomes	Readiness and Resilience Building Block(s)	Conceptual Link Between Short-Term Outcomes and Building Block(s)
Increased social interaction	• Social network	This building block refers to the availability of social connections to an individual. It is expected that increased social interaction within the unit helps build the strength of the connections among unit members.
Positive use of leisure time	• Coping strategies and skills • Involvement in activities	The coping strategies and skills building block refers to practices that help an individual adapt to stress. Our previous work (Meadows et al., 2019) highlighted ways in which leisure can be used as a type of coping strategy, in part by acting as a buffer to stress (see Caldwell, 2005; and Iwasaki et al., 2002). The involvement in activities building block refers to participation in leisure, community, and social activities (e.g., Sanderson and Brewer, 2017; Zolkoski and Bullock, 2012), which can support mental and behavioral health.
Opportunity to decompress	• Coping strategies and skills	This building block refers to practices that help an individual adapt to stress. The leisure coping literature has indicated that opportunities to relax and rejuvenate can serve as a coping strategy (Iwasaki, MacTavish, and MacKay, 2005).
Promotion of Air Force institutional values	• Peer group/unit values • Community/Air Force values	The peer group/unit values building block refers to group beliefs and attitudes tied to a social identify shared within the group (McFadden, Campbell, and Taylor, 2015; Morgan, Fletcher, and Sarkar, 2017). The community/Air Force values building block refers to shared community beliefs and commitments (e.g., Hopkins-Chadwick, 2006; and McGonigle et al., 2005). Both building blocks are directly relevant to the promotion of Air Force values.
Increased physical activity	• Physical health • Involvement in activities	The physical health building block refers to physical well-being, including physical activity and nutrition (Holland and Schmidt, 2015; Yousafzai, Rasheed, and Bhutta, 2013). Whether a UNITE event includes a component of physical activity is directly relevant to this building block. In addition, the involvement in activities building block refers to participation in informal or formal activities, including recreational activities (e.g., Chen and Kovacs, 2013; and Zolkoski and Bullock, 2012). In the building blocks model, involvement in activities is conceptualized as a building block that can contribute to the physical health building block, such as with group physical activities, which are common in UNITE events.

Evaluation Framework

In addition to summarizing the operation of the program, this logic model is the basis for our evaluation of UNITE, which includes both process (i.e., resources and inputs, activities, outputs) and outcome (i.e., short-term and intermediate outcomes) components. Note, however, that we do not extend the evaluation into a longer-term assessment of impacts (i.e., airman and unit readiness and resilience), given challenges to accessing data needed to measure these constructs.[9]

Our evaluation included a combination of quantitative and qualitative data sources, as well as perspectives from the variety of stakeholders involved in UNITE. Table 2.3 provides an overview of our analysis, noting the source of the data, type of data, and methods used in each analysis. It also denotes the perspective gained by each data source (e.g., airmen, unit commanders, C3s). Details are provided in the subsequent chapters and appendixes of this report.

Table 2.3. Overview of Data Sources and Methods

	Data Source(s)	Perspective	Data Type and Method
Process/implementation evaluation			
Resources and inputs	• C3 interviews • C3 email survey	• C3s	• Qualitative (structural and saturation coding)
	• AARs	• Unit commanders	
Activities	• C3 interviews • C3 email survey	• C3s	• Qualitative (structural and saturation coding)
	• AARs	• Unit commanders	• Qualitative (saturation coding)
	• UNITE post-participation survey	• Airmen	
Outputs	• AARs	• Unit commanders	
	• UNITE post-participation survey	• Airmen	• Qualitative (saturation coding) • Descriptive statistics
Outcome evaluation[a]			
Short-term outcomes	• UNITE post-participation survey	• Airmen	• Multivariate regression
	• C3 SharePoint database	• C3s	
Intermediate outcomes	• UNITE post-participation survey	• Airmen	• Path model
	• C3 SharePoint database	• C3s	

NOTE: Structural coding is a process in which codes are based on study objectives and interview questions and are intended to identify themes in the data. Saturation coding is a process in which development of new codes stops once no new information or themes are observed in the data. More details on methods can be found in Appendixes B, C, and D.
[a] The outcome evaluation also uses Air Force personnel data.

---

[9] We address possible options for an impact analysis of UNITE in Chapter 5.

Summary

In this chapter, we provided a brief overview of our examination of the cohesion literature, focusing on factors that predict or facilitate cohesion and how those factors overlap with our earlier work on the building blocks of readiness and resilience. During our review of predictors of cohesion, we found very few studies that discussed implementing and evaluating a program or initiative designed to improve group or unit cohesion. When we did find such studies, they were often methodologically weak and, at best, offered speculative evidence that such programs work. Obviously, despite our systematic review, we may have missed some cohesion intervention evaluations. In addition, some interventions may not have yet received any type of evaluation.

Despite the weak evidence base regarding cohesion interventions, it appears that the UNITE Initiative targets cohesion predictors (and readiness and resilience building blocks) that are associated with group cohesion, though perhaps not causally. Thus, we sought in the next phase of the study to understand (1) how UNITE was implemented and (2) what it was achieving. In the next two chapters, we review the results of our implementation evaluation (Chapter 3) and our outcome evaluation (Chapter 4) of UNITE.

Chapter 3. Process: How Is UNITE Implemented?

In this chapter, we rely on qualitative data to examine how UNITE is being implemented. This process evaluation can provide information that the Air Force can use to improve the efficiency (and, ultimately, the effectiveness) of UNITE. As outlined in the UNITE logic model (see Chapter 2), a process or implementation evaluation focuses on resources/inputs (e.g., funding, C3s), activities (e.g., planning and coordinating UNITE events), and outputs (e.g., satisfaction). With the process evaluation, we aimed to answer the following questions:

- How are UNITE events planned and executed?
- What obstacles do C3s face in planning and executing UNITE events?
- From the unit command perspective, what went well, and what areas are in need of improvement? What lessons learned have emerged during the first year of UNITE implementation? Are airmen satisfied with the UNITE events they have participated in?

The qualitative data used in the process evaluation come from various sources and groups of people. Our primary data collection methods focused on C3s. We conducted in-depth interviews with a sample of C3s who were selected to represent a variety of installations with respect to MAJCOM, installation size, remote or isolated status, and geography. These interviews focused on understanding C3 training and resources, the process of planning and executing a UNITE event, and implementation challenges. All C3s also were invited to participate in a brief email survey that took place approximately three to six months after the C3 interviews. This survey allowed us to obtain more up-to-date information about common C3 practices and obstacles experienced from a broader sample of C3s than those who participated in interviews.

We also had access to two secondary data sources that we analyzed for the process evaluation. First, AARs allowed commanders and POCs from units participating in UNITE events to provide information about their experiences. In addition, an open-ended survey item was included on the post-UNITE participation survey completed by UNITE participants. Although these secondary sources of data were not formally collected as part of the evaluation, we reviewed and analyzed them because they represented additional perspectives on UNITE implementation (i.e., from commanders, POCs, and participants).[10]

[10] Although the C3s invited to participate in interviews were selected to be representative of installations implementing UNITE, there may be selection effects among those who responded to the email survey. In addition, AAR data and open-ended survey data were not collected as formal sources of evaluation data, and the open-ended survey data are subject to concerns about selection effects (described further in Chapter 4 and Appendix C). In qualitative research with samples with limited generalizability, there often is a preference against reporting specific frequencies of codes or categories (Levitt et al., 2018; Maxwell, 2010; Neale, Miller, and West, 2014). Therefore, we have not reported frequencies or raw numbers when presenting qualitative findings throughout this chapter.

In addition to these qualitative data sources, we have quantitative survey data from UNITE participants, which allows us to measure their satisfaction with the UNITE event they attended. In the rest of this chapter, we review these data sources. Throughout this chapter, we provide summaries of key findings by data source, including tables with example quotes. The tables with quotes are most likely to be of interest to stakeholders involved in the execution of UNITE. At the end of the chapter, we synthesize results from across data sources and summarize the key findings as they relate to how closely the actual implementation of UNITE mirrors the ideal that is outlined in the program logic model. Methodological details for all sources of data, coding, and qualitative analyses are available in Appendix B.

Interviews with Community Cohesion Coordinators

In fall 2019, we interviewed a sample of C3s administering UNITE at a diverse set of Air Force installations. We conducted the interviews via telephone using a semistructured interview protocol, which meant we had a common set of starting questions, but follow-up questions varied according to interviewees' initial responses. Interview questions covered training and other resources available to C3s, UNITE dissemination strategies, processes related to working with squadrons to plan a UNITE activity, program implementation challenges, and recommendations for improvement. The interviews were documented by a dedicated notetaker and analyzed using a computer-assisted qualitative data analysis procedure, referred to as *coding*.

We invited 24 C3s to participate, of whom 22 took part in an interview. Note that C3s were selected to ensure that the interview sample was representative of the total C3 population in terms of their installations (e.g., inside the continental United States [CONUS] or outside the continental United States [OCONUS], MAJCOM, number of assigned personnel) and the local area (region), but it is possible the interview sample differs from the overall group of C3s in other ways, such as C3 individual demographics or C3 job performance. It is not clear how such differences would bias our findings. However, the lack of information on C3 job performance or installation-level UNITE success prevented us from identifying best practices or comparing practices of more-successful C3s and installations against less-successful ones.

In the sections that follow, we report interview findings related to the following topics:

- C3s' role in planning and executing UNITE events
- UNITE events held by units
- C3 resources
- challenges related to UNITE implementation.

To convey the nature of our evidence and illustrate key points, we include example quotes from the interviews.

Community Cohesion Coordinator Role in Planning and Executing UNITE Events

Our interview protocol included a series of open-ended questions that were intended to gather information on how C3s were implementing UNITE at their locations. A large portion of the interview was dedicated to understanding how C3s work with units to bring about UNITE events and how that practice varies across units and installations.

The bulk of a C3's unit-focused interactions tended to be with the unit's UNITE POC, who, in turn, interfaced with their unit commander as needed, such as for a sign-off on an event proposal. Comments about C3-POC interactions occurred in about half of our interviews. As the following excerpts illustrate, these types of interactions covered the program goals, parameters, and requirements and were intended to set the stage for planning specific activities:

- We discuss what falls within . . . the program; the goal of the program—revitalizing your squadron; we discuss the requirements, the unit cohesion or team-building component; we discuss the allocations, how it's $17.50 per person for nonfood-related costs and $5 per person for food-related costs.
- [T]ypically, [after I get an email,] I pick up the phone immediately and I start building that one-on-one relationship. And try as much as possible to also set up an in-person discussion and meeting—this is what all the forms look like, this is what we need prior, and now let's start talking about our event, and here's your available funding, what can we do that's within this.
- I send them an email briefing them on all the forms required—the event proposal, AAR—and the brochure of activities. So there is some consistency among the program. I also send the DoD ID [U.S. Department of Defense identification number] form.

Those remarks are also examples of a less-structured, informal training approach for POCs: information about UNITE conveyed to individual POCs via a telephone conversation and/or email. However, some interviews included references to structured POC training, typically for a group of POCs but sometimes delivered one-on-one if a POC could not attend a group session. For example:

> We also had a face-to-face training for all the POCs, and if they didn't make it, I went squadron to squadron to train the POCs. I scheduled time to train the staff onsite. Here is one thing we have learned: If they can't come to us, we need to go to them. We want to make it easy for them. There are [NUMBER] squadrons at [INSTALLATION] that I physically went to. We lay the instructions out, have briefings on the SharePoint site they [POCs] can download, and send out a monthly or as-needed update to all the POCs.

One C3 explained that she adapted AFSVC C3 training content to suit this purpose:

> [I]t was pretty much taken from everything that we had gotten from the Air Force Services Center when we had our initial training. So a lot of slides and everything that we were training on [became my training for the POCs]. It was more condensed because I [created] more of a daylong training. So it condensed the information that they needed to have. And as we received new updated information from the agency, that in turn adjusted some of our training materials.

We also reviewed the interviews for interactions that occurred in the process of planning a specific UNITE event. To initiate that process, C3s routinely made event menus or catalogs available to POCs. Just more than half of the C3s interviewed discussed having something like a catalog, menu, brochure, or list of event options to aid units in event planning. As the following comments indicate, the event compendium (in some cases) was compiled using a lot of research and/or networking by the C3:

- I sit down and have a continuity book—a bank of resources within a two-hour radius from [INSTALLATION] along with our FSS resources. So we have Top Golf, paintball, pool party—I have that whole list set up for them.
- I give the POCs a list of all the businesses and people at those businesses I've contacted that they should talk to.

A few C3s were less precise in their language, referring to providing POCs with ideas but not specifying if that was orally or via a printed reference. One of those C3s mentioned not having time to develop a catalog:

- I give them ideas on things I've looked into locally and have already made pricing arrangements on. If they have ideas, I'll be more than happy to work with them and develop the program to meet their needs.
- I do help them, but I do not have a catalog for them. I have not had time to do that.

According to the UNITE CONOPS, one of the goals of the program is to maximize the use of FSS facilities, but we found that other factors (e.g., unit preferences) might affect whether C3s adhere to this goal. For example, one C3 said,

> My immediate supervisor . . . asks me why I am networking with people, why I am going somewhere else [off base] when there is bowling on base. He doesn't understand it's supposed to be the unit's prerogative. I can't dictate where people want to spend their money. He wanted them to spend money on base.

Other C3s expressed views more consistent with the guidance:

- I have a brochure with all of the on-base options because our goal is to push the use of FSS facilities.
- We're all about keeping money within FSS, that's really important. I push "let's do an event at the pool, let's do an event at the club" because it's easy to get there and it keeps money within our program. But bowling, pool party, club, can get a little old; you want to think outside the box. You can do that in your outside activities [i.e., off base].

The last two remarks suggest that C3s were mindful that although the Air Force prefers FSS programs to be tapped for UNITE events, FSS facilities have their limitations, and some units were keen to hold an event off-base regardless of C3 efforts to persuade them otherwise. Thus, some C3s dedicated time to developing not only FSS-based ready-to-execute activities (RTEs) but also events that tap into local community resources. As the previous comments about event catalogs indicate, this practice meant that C3s at some installations had conversations with local vendors to identify UNITE-appropriate opportunities available off base and to negotiate pricing arrangements, at times obtaining discounted rates. One C3 told us that C3s were *supposed* to be comprehensive in this way:

> We [C3s] are all told to create a menu of program ideas in the local community on and off base, and that is great; it gives us some ideas they [units] may not have thought of.

Overall, there was clearly a variety of views within our interview sample regarding the extent to which C3s could or should promote off-base event possibilities, as well as variation in C3 efforts to develop and publicize to units the availability of affordable, UNITE-appropriate off-base RTEs.

The interviews suggested that POCs tended to come to C3s with well-formed event ideas, possibly because of efforts by C3s to provide units with a variety of RTEs. Yet there were still some POCs who could not decide what their unit should do and some who had an idea in mind that was not appropriate, at least as initially conceived, for UNITE funding. In those circumstances, C3s worked closely with a POC to develop an idea or shape an idea so that it was suitable for a UNITE event:

- Some of the squadrons work together within the same flight and figure out what to do. I help them put it into the UNITE program. For example, they want to do a motorcycle ride. How is that team cohesion? I help them make it [their idea] into an approved UNITE event.
- I had a squadron two days ago that reached out to me about a place off base—a pumpkin farm. The squadron is doing some team-building there, and they reached out to me to see if I could purchase pumpkins and the way she wrote the email to me is "We want to see if we can use UNITE funds to purchase pumpkins for the event." My initial thought was no, that is not recreation-related . . . if it's just pumpkins I can't pay for it. So I had to really brainstorm and talk it through . . . turns out they want the pumpkins to do team pumpkin-carving contests with squadron leaders, so that said I was able to bill those pumpkins because it was team-building related. You have to be creative and not necessarily say no. You have to find ways to be able to say yes.

Developing event ideas was an important way that C3s worked with units, but our interviews showed it was not the only basis for C3 interactions. For example, C3s told us about warm transfers in which they connected unit POCs with facilities managers, both on and off base:

- If they want to do a picnic I put them in touch with equipment rentals.
- I give the POCs a list of all the businesses and people at those businesses I've contacted that they should talk to. The POCs call me or send me an email, and I tell them who to go talk to, and I send the businesses an email to let them know that an email from a POC is coming. Most of the places they go to, I have already talked to before.

C3s also helped with unit event planning by submitting the required approval paperwork (an event proposal form) to AFSVC for approval. This was an essential but typically simple and straightforward way that C3s worked with their units. In contrast, making event purchases on units' behalf was rather demanding for the C3s at times. Because UNITE expenses need to be paid for via a government purchase card, C3s found themselves busy supporting units by paying for their event's food and other costs. Sometimes this could be handled quickly because the facility was on base or because remote payment was possible. For example, one C3 told us, "If I'm paying for a bowling event but I'm offline, they have no problem taking a payment over the phone." Other times, this meant more legwork for the C3s:

> Picking up the food has been a big role of mine. When you order food from some place—Texas Roadhouse, Olive Garden—they want to be paid on the day of the event. I go to the food location to pick up and pay.

One C3 brought up several problems related to this need for the C3 to personally pay for all UNITE event expenses and raised a concern about whether paying for goods or services over the telephone was a violation of Air Force policy:

> My role is to provide funding. It's a bit frustrating, but if I have to go to the commissary five times to meet five different groups, I do. Sometimes it [the C3 providing funding] can be done over the phone. It's somewhat a violation of key card [policies] though, because I'm supposed to have a physical receipt and sometimes it's hard to get that from the POC. It's frustrating for the POC too because they have to work around my schedule [for an in-person food purchase].

The final aspect of event planning that we covered in the interviews was its culmination: a UNITE event executed by a unit. On the day of the event, there are no requirements or guidance as to what C3s should do beyond making any outstanding payments. We found that C3s vary in the roles that they play. Some C3s indicated they typically do not attend a unit's UNITE event:

- I do not attend an event unless they invite me personally because I don't want to take any attention away from the POC. I see it more as I want those POCs to shine because they are normally a younger ranking airman.
- But I don't go to the event or take pictures of their events. That's up to the POCs on their own. They are executing it; I am only purchasing, submitting for the approvals, collecting the information from the AAR, and putting that [information] into our systems.

Others occasionally attended events, usually playing a limited role when they were there, such as making sure the event was going as planned:

- I don't go to all of the events. I try to get out there to most if I can . . . Sometimes I like to pop in and see my work, so to speak—to see if people are having a good time and if there are any hiccups and ways I can improve.
- If they are on the installation, I go to make sure the setup is going okay. If they rent items from Outdoor Recreation, I will make sure they got what they needed, and I will take a picture. And once it is going smoothly, I got out of their hair and let them [POCs] roll with the event. The off-base events, I didn't attend any of those.

Perhaps in recognition of the lack of guidance in this regard, a small number of interviewees advocated for C3s to play a greater role in UNITE events, and they tried to do so as their schedules permitted:

- In a small base, I can go in to help at an event usually. Having [had] the MRT [Master Resilience Training] and resiliency training and being able to look at things [in person] at events, I think it would be good if the C3s can have these one-on-ones. For me, that's been helpful . . . You don't have to, but there are some sections in a squadron that are smaller and may need someone's help with team-building and ideas . . . I can talk about resiliency to the units without it being a training.
- The day of, if I can attend, I will attend the event. I put it upon myself if I'm there to sign people in, because it's easier for me to get the DoD ID numbers that way. If I have to wait on [the POC to provide the information,] it takes too long. At the entrance, I'll immediately get their DoD ID number. Then I'll take pictures, make sure everything's set up and ready to go and hang out for a little while, and once everyone has come in, I'll thank them from the UNITE program. If it's a crazy interesting event, I'll hang out for a while. If it's bowling, I'll stay for a little while and then leave. I have shown up to most of the weekend events because I want to have a presence of why you're here and why it's important we're doing this, so I give my spiel at the beginning of each event.

UNITE Events Held by Units

In regard to what the units do, the C3 interviews offer limited but still-helpful insights about whether units are using FSS facilities for their UNITE events and why or why not. Some interviewees discussed the split of off-base and on-base events, and their comments ranged from almost all on base, to a relatively even split, to the majority off base. Their remarks suggest that the local context, available time, and prior UNITE experience might influence event location. As the following comments indicate, the weather or season and the availability of offerings on base were important contextual factors that shaped unit event selection:

- I would say in the summertime there are some events on base. Since the summer season has been over, the events have been 100 percent off base.
- There are not any military resources on [INSTALLATION]. One time, I coordinated a fun run at the gym, and, so far, that has been the one FSS event. Other than the gym, there is nothing else [on base].

In addition, a small number of C3s suggested that POCs and units might look beyond the base more often as those groups gain experience with UNITE. One of the C3s also felt that event location was influenced by the time allotted for the event:

> They are starting to venture out a little more now that we are getting into Phase Two of the program . . . A lot of folks have done their first event and now they've learned . . . It also depends on how much time they have for their event. Some only have a lunch hour and can't go out of base. Others do their event in after hours and do more in town.

Overall, off-base event choices seem to provide distinct experiences that are not offered by FSS programs. Top Golf and escape rooms were repeatedly mentioned. Other off-base options cited during the interviews include some operated by commercial vendors, such as Paint and Sip, paintball, laser tag, climbing wall or climbing gym, ax throwing, ziplining, and shooting ranges. For example,

> if the unit develops programs, [that means] they take time to go out and look at local resources around to see what they have available. Some will choose a thrill park or paintball or something more ready-to-execute.

However, looking beyond FSS program offerings does not necessarily mean turning to commercial vendors. Not only did some non-FSS events take place at public recreation areas, such as parks, beaches, and trails, but units at some locations also tapped into nearby U.S. Department of Defense (DoD)-run facilities:

- And it's a big community for [STATE] [in terms of] what can we do—escape rooms, movie theaters, parks, trails, bike rides, zoo. It's really just starting to talk to them and see how we can make it happen from the get-go.
- When you say off base, it doesn't mean we are not utilizing other [military] bases . . . One event was held on a large beach on one of the bases. They also had an event in an indoor shooting range on another base.

One type of off-base event that did not seem to be very popular, at least not yet, was volunteer events. We did not expressly ask all C3s about volunteer events, but the topic came up in multiple interviews. C3s mentioned units planning events to support their local community of individuals experiencing homelessness, and one C3 expressly noted including volunteer options in her event catalog. Another C3 discussed the value of volunteer events as a kind of "force multiplier" that could enable a unit to have more events in a calendar year:

> If they want to do one amazing event per year—going back to [SPECIFIC COSTLY EVENT], that was amazing. People said it was the best they had been to. Maybe next year they'll do it again. But thinking about, how do we keep that glue for the rest of the year when you forgot that person's name already a month later who's on the other hall . . . I don't think more funds [would make a big difference], because $17.50 per person is good. You just don't want to become money bags, that's not the point of the program, and that's where the volunteering can come in.

On the topic of funding, a few C3s observed that the amount and type of funding available to units affected their interest in volunteering. When they only had NAF, they looked at volunteer events, but when APF was available, units wanted to spend it:

- Some bases only received NAF dollars [for food], and they did more volunteer events since they did not have APF dollars to do another event. Since we had both sets of money, they [units] haven't really done the volunteer events.
- I've been pushing the volunteer events, and all I keep getting is that "we can volunteer on our own." I keep trying to push them to volunteer, but that's

> been a hard one for me to do so far. They got money, they want to spend the money because they don't have to do a fundraiser to get the money.

Community Cohesion Coordinator Resources

Initial Training

C3s were asked to describe the training that they received for their role. At the time they were interviewed, the majority of interviewees had attended the initial in-person training at AFSVC. Most participants described the training as useful, noting that it covered several topics that were valuable when starting out in the C3 role:

- The training was excellent. We stayed in San Antonio for a week, and the training addressed everything, including things like how to hold an event, about options like the golf courses and Outdoor Rec, and told us how all of the programs work together and how the C3s support the UNITE program.
- It was very useful. A lot of stuff we discussed in the training, I was able to apply to the job. Being creative, thinking outside the box, budgeting, giving commanders ideas on programs we might want to implement.

In addition to describing the substantive topics covered by the training, interviewees mentioned the importance of having opportunities to practice aspects of their roles during the training. For example, one interviewee highlighted role-plays, during which C3s practiced meeting a commander or squadron leader to explain the UNITE initiative. Several C3s also reported that networking with other C3s was one of the most valuable aspects of the training:

> I think getting to know my team members was essential. I think if you don't have face-to-face interaction with people you're going to be working with so closely who are 1,000s of miles away, potentially, you don't have that connection so much. So I think building that team dynamic was really important for us to all be out there and get that.

Although perceptions of the training were generally positive, interviewees also noted certain limitations of the training. C3s come from a variety of backgrounds; many have experience working in FSS or MWR programs, whereas others have a marketing background. Interviewees noted perceptions of the level of detail provided in the training may have been different for these groups. Those with prior FSS or MWR experience suggested that it may not be necessary to spend so much time during the training to introduce the programs that may be available on installations, such as bowling centers or outdoor recreation, because they were already familiar with these programs. But C3s from other backgrounds were less familiar with these programs, and found the amount of new information overwhelming at times:

> It was also a lot of information, and I felt like I was drinking out of a water hose. I was one of the few who did not have an FSS background, while a lot of others had backgrounds from either Outdoor Rec or MWR. It was hard for me to wrap my head around it all.

Regardless of background, though, many C3s stated that it would be valuable to have refresher trainings for C3s who have been in the position for a while. They noted that refresher trainings could provide an opportunity to update C3s on any new policies or procedures and would give more-established C3s the opportunity to share effective practices for planning events.

Other Resources

We asked C3s about the resources that they turned to for assistance after completing their initial in-person training. We learned that those resources include three types of Air Force personnel (other C3s, AFSVC personnel, and other installation-based personnel), off-base resources, and technology-based resources. Although we did not ask the interviewees to evaluate the resources they identified, they offered value judgments without prompting in some cases. In the next sections, we discuss the resources mentioned in the interviews and present notable C3 observations about the resources that were considered to be useful or lacking in some way. Additional findings related to resource shortcomings are covered in the section on challenges to UNITE implementation in this chapter.

The most-frequently cited resource was other C3s, with several interviewees mentioning interactions with other C3s to share tools and information, such as event ideas:

- I know [OTHER C3] at [OTHER INSTALLATION]—me and her bounce a lot of ideas off each other. I planned a [UNIQUE EVENT] and posted on Facebook, and people asked a lot of questions. I got UNITE signs made and people asked me where I got those, and I got folks a good deal from the company that I used.
- I get help from my colleagues [other C3s] as far as creating the spreadsheet and sharing documents and ideas and information on the Blackboard system or calling or emailing them about processes and whatnot.

The interviews suggest that there is a strong C3 network spanning Air Force installations, with C3s routinely tapping into each other's experiences to carry out their work. The C3s also mentioned turning to UNITE personnel based at AFSVC when they had questions. Those comments were fewer and less detailed in nature, such as "Over in San Antonio, [community program and unit cohesion lead coordinators] are a great help."

The third group of Air Force personnel, those based at the C3's installation, include a variety of personnel, such as leaders of other FSS programs, FSS flight chiefs, and marketing department personnel. As the comments that follow indicate, together these individuals helped C3s promote and execute UNITE:

- In the FSS flight I have an amazing team to network with. I have the Outdoor Rec [Program], ITT [Information, Tickets, and Travel], the golf course, and bowling alley. And my leadership chain has a wealth of experience for programming and operation. And any support needs, of course I have my commander and I have different . . . squadrons, and I have quite a variety [of resources] when it comes to getting help.

- I am working with the marketing department to develop a web-based sign-up system like the one at the Air Force Academy.

One C3 mentioned looking to the local military spouse community as a source of UNITE support as well:

> I've been able to connect with spouses that have unique talents or starting their own little business. Even though we have limited facilities here, I can bring in a painting instructor to do a class for the CDC [Child Development Center] as a UNITE program. I [meet] a spouse that does cake decorating, and we can do cake decorating as a UNITE program.

The majority of C3s interviewed also mentioned relying on off-base resources but said they did so for somewhat different reasons than their reasons for working with Air Force–based resources. Connections with off-base resources, such as vendors, allowed C3s to understand the various offerings in the communities. These connections also helped them build networks and stay current on potential offerings outside the base, thereby informing the development of the event catalogues mentioned earlier. One C3 said,

> I kind of check out a lot of different things locally in the area. I have checked with our local Chamber of Commerce for the communities that are close to our base, the main one being [CITY], but we live in an area where there are three or four different communities around. There are a number of publications that come out on a regular basis to see what's going on in the local area. I take a look at a couple of trade magazines to get information about other things that are going on . . . Looking at other parks and recreation departments, I look and see what they're doing with some of their programming and stuff like that, looking to see where a company can handle either small or large groups. We've had a couple of escape rooms and a couple of virtual reality places that have opened in the area recently, so those are new things that have come about in our community, so I am trying to stay on top of those things as well.

The last set of resources we heard about during interviews were technology-based: Blackboard, Defense Collaboration Service (DCS), a Facebook group, search engines, and websites. The first three are Air Force–managed technology tools, as follows:

- Blackboard is a collaborative educational platform focused on UNITE that is centrally managed, with information pushed out to C3s and C3s contributing to it using a bulletin-board format.
- DCS is a virtual web-based meeting platform used primarily for monthly meetings between UNITE AFSVC personnel and the C3s.
- The Facebook group is a newsfeed-based private discussion forum for C3s to chat informally, post questions, and share their experiences.

Internet resources mentioned by C3s include the Google search engine, vendor websites, and other websites maintained by public agencies or private organizations. These resources assisted the C3s in connecting with other resources, sharing information and ideas, finding events, and obtaining training. Table 3.1 features illustrative quotes from the interviews that depict how C3s rely on these resources for these purposes. In some cases, a single resource serves multiple purposes (e.g., the C3 Facebook group); in other cases, C3s draw from multiple resources to address a single need.

Table 3.1. Examples of Community Cohesion Coordinator Technology-Based Resources, by Purpose

Technology Purpose	Example Quotes
Sharing information and ideas	• As the year progressed, more of those answers came down either through email or through the online Blackboard system or through the teleconferences. • We also have a Blackboard page where I can look at what other people have done. It's a good resource where we talk about problems we've had. • We have a Facebook group chat. Once we do events, we'll post online and bounce ideas off each other.
Finding events	• I just did a Google search of different things in the area. • Usually I hear about them on the news and websites . . . The local downtown programs are listed on the local website and the events are mostly repetitive. • I check out the C3 group Facebook page to see what others are doing and see if there is something similar locally. • We have our closed Facebook group. [We are] posting pictures of our events and if someone sees something that that looks cool, they ask "What did you do?" So, we can then offer that at our base to offer to squadrons, for something they might like to do.
Obtaining training	• They regularly schedule meetings via DCS, which is basically a computer teleconference that you cannot see each other in. It's a secure unclassified channel for the military . . . On DCS, they provide briefings and updates and there is an opportunity for us to ask and send in questions early to discuss it. • There's also a lot of great online education. We do monthly videoconferences with [Community Program and Unit Cohesion Lead Coordinators] and my [C3] peers across the world so we're able to keep abreast of things . . . best practices, things that work, things that don't. • We do have a conference call and training once a month. They will give us example of requests that are appropriate or inappropriate and discuss them.

SOURCE: Interviews with C3s in 2019.

Perceptions of Resource Usefulness

C3s who said they relied on AFSVC personnel or installation-based personnel tended to speak of them in favorable terms. For example, the central UNITE leaders were described as a "great help" and "quick in responding," and local personnel were referred to with such terms as "wonderful leadership," "amazing," and "helpful." Given the limited amount of time we had with C3s, we opted not to probe on positive views; rather, we tried to learn more about the resources with mixed reviews to identify room for improvement. This latter category included other C3s, the Facebook group, and Blackboard. As the following comment indicates, some interviewees felt that other C3s may not be a good source of support because of the different context they were operating in or because their way of doing things may not necessarily be effective:

> To be honest, I don't reach out that much. I don't feel I have the same struggles, so even if I called [OTHER INSTALLATION] or [MAJCOM], they are not going to have the same issues. Apart from venting to each other, I don't know

> how helpful we are to each other because each of the installations is so different. Here at [INSTALLATION] we have low manning [overall], including in our food facilities, so I almost never request food for an event because I am gambling so much that it may never get done.

The reluctance to rely on other C3s also meant that the Facebook group and Blackboard platform were potentially less useful because they were conduits for C3s to share their knowledge and experiences. Other C3s also mentioned that Blackboard has limits because of its impersonal, virtual nature:

> We have the Blackboard but it's not the same as in person. I personally feel a little more vulnerable on Blackboard because once you type it and send it, it's permanent. You can't spread your emotion through it. I would rather talk face-to-face.

Although the quality of answers received via Blackboard may not be influenced by the perceived impersonal nature, it is possible that Blackboard is less effective than it could be because some C3s are reluctant to use the virtual platform to ask questions or share their experiences.

Challenges Related to UNITE Implementation

During the interviews, we asked C3s to tell us about any challenges or barriers they experienced in their efforts to implement UNITE. We coded their responses to that open-ended question and any factors the interviewees cited as an impediment at any point during the discussion. The challenges fall into three broad categories: UNITE-related, unit-related, and installation-related, which we describe in the following sections.

Challenges Related to the UNITE Initiative Itself

The majority of interviewees described challenges related to how UNITE has been rolled out. Four challenges pertain to overarching UNITE issues: governance, funding, the C3 SharePoint site, and the requirement to collect DoD ID numbers from participating airmen; two challenges are local in nature, related to local resources for the C3 and local FSS leadership.

Table 3.2 lists the overarching challenges and example quotes from the C3 interviews. The first set of such challenges pertain to UNITE governance: formal guidance, process standardization, and organization structure, with several C3s expressing concerns about at least one of these governance elements during their interview. Guidance is an important part of program governance; in the case of UNITE, C3s were critical of the lack of useful guidance and how guidance had evolved since the program's inception. They focused on guidance related to acceptable uses of UNITE funding and designated sources of marketing support. One C3 lamented the lack of an Air Force Instruction (AFI) in particular. Process standardization was another aspect of governance that C3s felt was lacking in some ways. Interviewees offered examples of C3s developing their own marketing materials and forms for event proposals, event tracking, and funds tracking. C3s viewed this lack of standardization as a challenge because it meant more work for them, but it could also be a problem for UNITE as an initiative if greater consistency is desired in terms of processes and procedures. Finally, challenges with structure were mentioned by a small number of interviewees. Specifically, these C3s felt that their (and UNITE's) placement within the FSS flight limited their ability to implement UNITE because it was harder for them to reach wing commanders. Being at the wing level was seen as a source of authority. One interviewee also noted that the Resiliency Program was at the wing level and raised the question of whether UNITE should be at the same level or fall under that program's purview.

C3s also discussed challenges pertaining to both the amount of funding available (for food in particular) and the timing of when funding was received. Interviewees who raised this issue felt that $5 per person for food was insufficient at times, especially when units opted to hold a UNITE event off base. Comments about funding timing were related to either the delay in receiving funds during UNITE's first year or the different timelines for using APF and NAF (calendar year and fiscal year, respectively). According to the C3s we interviewed, these timing issues influenced when units held their events. The funding delay meant many units held events during a relatively narrow time frame, and some units opted to wait until they had both NAF and APF available for their events. The different funding timelines also created tracking-related challenges for both the C3s and the units, generating more work for the former and potentially demotivating the latter.

The last two challenges in Table 3.2 were brought up by only a few interviewees but could affect a larger portion of the C3 population. The first relates to the C3 SharePoint site, which serves as a platform for collecting AAR inputs and other event-related information. Several C3s discussed problems related to SharePoint that they felt made their work more difficult. One had trouble learning how to use SharePoint in general, while others focused on problems with data entry and printing or otherwise viewing the data that they inputted. Second, C3s noted that meeting the requirement to collect DoD ID numbers from UNITE participants for the post-participation surveys could be difficult. Specifically, C3s talked about airmen's reluctance at times to provide their DoD IDs. They noted that obtaining airmen's DoD IDs was especially problematic during larger events.

Table 3.2. Perceived Challenges Related to UNITE: Overarching Issues

Challenge	Example Quotes
Governance	Formal guidance: • The fact that we can't buy equipment [to use UNITE funds in this way] has evolved. We originally could, and now we can't. • Another challenge is that there is no AFI directive. It would be nice for them to send something out [that states that] we cannot use UNITE money for shirts, commander's call, etc. It would be nice to have something that says there are definite things you cannot use UNITE money for. Process standardization: • There is not really a standard welcome letter to explain the UNITE program, and it is up to the individual C3 to come up with one. • When the program was first rolled out, everyone was creating their own forms and doing their best to create what works for their installation. I wouldn't be surprised if a good percentage [of] each of us has created our own sheets. Organization structure: • In the Community Services Flight, it is hard for us [C3s] to go up the squadron command. If we were at the wing level, we can tell the commanders to participate. • I think there are some issues with organization. I'm part of the Comm Services Flight . . . and at the wing level, the Resiliency Program specialist thinks this program [UNITE] should be under her. I don't agree or disagree; I'm just stating that is her view.
Funding	Amount: • $5 meals. $17.50 per person for a program is reasonable, but $5 a meal is insufficient. $7.50 to $10 would help out. I know the corporate response is that the money is not intended to cover all program costs, [instead] it's to offset costs. I get that, but if the majority of the folks aren't able to execute all the money and money is left over, reassess it and increase the allotment for food. Timing: • This year, the timing of when the funding became available [was a bit of a problem] . . . When we first received the funding back in April, that really didn't give us a lot of time to be able to get everybody trained up and get them used to the process and then get out there and start planning their events, and implementing their events, especially trying to get it done by the end of September. • Also with funding, if every single person was using their food money, it can be a tracking nightmare . . . We have to track money across quarters for the calendar year and the fiscal year. Some squadrons still have food money from the last fiscal year and some don't.
SharePoint site	• The hardest part was learning the SharePoint on my own. I've learned features through trial and error and it wasn't went over in the training. It's not a software that is commonly used. • We're using a SharePoint site to input the request, and there's no way to easily print out what we're doing for records or to view it when we're out in the field. Going back to the after action report, DOD ID numbers—it's just very difficult. • When we put in the DoD number, we put it in a ten-digit format with a comma and a space. If I put in 25 ten-digit DoD numbers, the system says I have put in 30 . . . After I put in the numbers, the system says 30 people showed up, and I have to manually correct it to say 25 people showed up. It's not a big deal with few people, but for an FSS picnic, it involved 350 people.
DoD ID requirement	• That's a struggle. A lot of [airmen] don't want to give the DOD IDs because they don't want to do the survey portion. I explain the survey portion, how many questions, show them on a piece of paper, and some just won't do it because they don't trust it. • For a picnic, it is harder to get those numbers. You will have 20 people say they will attend, but you can have 100 people show up. So with a squadron picnic, they are a little reluctant to collect those numbers because it seems like a meeting rather than having fun.

SOURCE: Interviews with C3s in 2019.

Turning to installation-level challenges related to UNITE, C3s mentioned issues related to a lack of local resources and support from FSS leadership. Table 3.3 includes example comments about both issues. Half of the interviewees talked about challenges stemming from insufficient manpower, marketing, or equipment dedicated to UNITE at their installation. They said they felt overwhelmed at times by all the work on their plates, including shopping to support multiple UNITE activities occurring close together, trying to attend events in person, and completing paperwork related to purchasing card (P-card) transactions and other UNITE administrative requirements. C3s suggested this manpower issue could be a problem especially at larger installations with many squadrons to support or as UNITE expanded to include Reserve Component personnel. A lack of local marketing resources was seen as hindering UNITE implementation as well. This included not only media advertising but also marketing assets and collateral, such as UNITE shirts, pens, posters, and brochures. As a small number of interviewees explained, C3s do not always have their own budget for UNITE marketing and rely on personal funds to purchase such materials (see Table 3.3). And as noted earlier, guidance about marketing-related responsibilities was perceived as lacking. Another C3 noted that this lack of resources was in apparent direct contrast to marketing available to support other FSS programs—examples of which we noted in the earlier section on C3 resources. Finally, some C3s mentioned that mobility-related equipment might present an implementation challenge. Specifically, the need to use a personal cell phone and personal vehicle for UNITE transactions, errands, and event attendance was seen as potentially wearing C3s down.

Although we reported earlier in this chapter that some C3s regarded their FSS chief as a helpful resource, a small number of interviewees indicated that their immediate leadership made it difficult for them to implement UNITE at their locations. One interviewee explained that her leadership wanted her to work on other FSS tasks and delayed her efforts to initiate UNITE, while another mentioned that her leadership was reluctant to have UNITE funds used for off-base events, even if that was a unit's preference. These C3s were also among those who felt that placing UNITE in the FSS flight versus at the wing level (discussed as part of the governance challenges) was problematic.

Table 3.3. Perceived Challenges Related to UNITE: Installation-Level Issues

Challenge	Example Quotes
Local C3 resources	Manpower: • I feel like there are a lot of squadrons eligible here, and it is hard for me to get to all of them. I find myself always talking to the same five POCs. I wish there was another one of me keeping up with the paperwork. When we use our P-cards we need to get a prior approval, form 43. It is not super difficult, but I have to send it to three different folks . . . [NUMBER] squadrons are a lot for me to focus on. • [W]e are available to everybody all the time, and sometimes especially at the end of the FY [fiscal year] there were so many people that wanted to have an event. So I had so many events that I felt like I was being torn in all directions. If there was help [additional manpower] that would be greatly needed maybe at the larger installations vs. the smaller installations. Marketing: • On some installations, Marketing does a lot for the program, but that's not the case for me. I got jealous seeing everybody on Facebook [the C3 page] with marketing things [assets]. Here, I paid for it myself, and the deputy paid me after learning about it. I am used to programs not having money, but it's just sad that I don't have the support of my marketing team. • There is really no written word, and there is no marketing of this program. The standard marketing for something like ITT or at the youth center would be saying what's happening. Other force [FSS] programs get standard marketing and advertising in print and e-print, but this base has not advanced to that with UNITE. I know other bases have greatly wrapped UNITE into their standard marketing and get all kinds of support. Equipment: • For me in the position being new and limited real estate—I have changed offices six times since starting this position—so I just started using my cell phone. I get email, texts, and phone calls on my personal phone . . . I am not always in my office, so if I am out purchasing for an event, how am I going to answer for my phone calls? • I have also seen on Blackboard a lot of us use our own personal vehicles to scout out the programs. Realistically, we won't ever get a vehicle for the UNITE program, but if there is not a government vehicle available, there should be some room to claim mileage. At some point, people will get tired of paying for their gas.
UNITE leadership: FSS level	• It was not a priority for the FSS here until the very last minute. I was waiting for the green light [from my FSS leadership] and did not get it until the very end. • We should report to the wing commander versus to the FSS flight chief, who has his own agenda. For example, I was trying to do things at the golf course, but I was told not to use it because the flight chief has hoped to close it down. Other C3s might not have that problem.

SOURCE: Interviews with C3s in 2019.

Challenges Related to the Units for Which UNITE Is Intended

The majority of C3s we interviewed identified unit personnel (unit leadership or POCs) as barriers to implementing UNITE. Illustrative comments about these types of challenges are listed in Table 3.4. Unit leaders, such as squadron commanders, were regarded as a barrier because of their lack of involvement in UNITE. Some C3s suggested this was due to their frequent turnover and because many do not regard UNITE as a priority in light of other demands on unit time. Another unit leadership challenge pertained to lack of awareness of UNITE and C3s' difficulties in getting the word out to units. The C3s who cited this obstacle expressed concern that their marketing efforts were not reaching unit leaders. The newness of UNITE and the newness of some officers to squadron command exacerbated C3s' marketing challenges.

The personnel designated by commanders to plan UNITE events also were regarded at times as a barrier to implementing UNITE. Half of the C3s we spoke with described unit POC-related problems, including not participating in UNITE training, turnover, and (most notably) the "teeth-pulling" and "babysitting" required to get AAR inputs and other event closeout-related materials, such as photos.

Table 3.4. Perceived Challenges Related to Unit Personnel

Challenge	Example Quotes
Unit leadership involvement	• I think just with the amount of squadron leadership changes, the turnover, getting all of them back onboard and utilizing the program has been a challenge. • I don't feel commanders are very involved. There are a couple of commanders who are like "this is great," but in general they are just like, "It's just a summer picnic and we have a mission to do."
Unit leadership awareness	• We had one squadron that contacted us the day after we were supposed to have funds spent by, and that was one of the ten [that hadn't participated]. And he was a MSgt [master sergeant] and he said that he had never even heard of the program. So that would be one thing I'm concerned about going forward, how do we best channel the information to the people that are going to act on it because it could get to one person and stop right there. • A lot of them are first-time commanders, and the program being new doesn't help since nobody knows about it. If we have this conversation in five years, I think everyone will know about the UNITE program. Commanders coming in from the Ops [Operations] side don't know anything about it.
Unit POC	• My role after the event is that of a babysitter. I say that because it doesn't matter how many times you need to tell the POCs, "I need to close out an event; I need the pictures and the AAR," [I still have trouble getting the information]. I had one squadron with a guy who planned the event going TDY [temporary duty] and leaving another guy in charge of the event the day of. I spent a month chasing him down to get what I needed to close out the event. There is no continuity whatsoever. • Some squadrons change POCs after the event. I have to explain everything again, and when the new POCs put in their request forms, you can tell they are not informed.

SOURCE: Interviews with C3s in 2019.

Challenges Related to the Installation

A final set of barriers, summarized in Table 3.5, related to the installation and thus may be less amenable to changes. These barriers included Air Force operational practices and aspects of the installation and the community in which it is located. Some C3s discussed operational practices as a challenge, referring generally to mission requirements and operations tempo or describing specific aspects of the installation's mission (e.g., intensity of exercise schedule during summer months, inspection schedules) that made it harder to implement UNITE at their location.

Half of the C3s discussed location-related challenges, which were more varied: Some were about installation characteristics, and others were about the local area. C3 comments about installation-related challenges touched on the capacity and quality of FSS programs and facilities, as well as on characteristics of civilian personnel. C3s also noted the challenges related to carrying out UNITE in remote, isolated, or OCONUS locations. These locations tended to have few event options available off the installation, thereby limiting units' options, and the OCONUS challenges varied by location. Depending on the country, challenges could include currency exchange issues, security restrictions, and language barriers (when trying to plan events off base).

Table 3.5. Perceived Challenges Related to the Installation

Challenge	Example Quotes
Air Force operational practices	• I think just the nature of the base and mission, each is different, and I think sometimes it falls off folks' radars because of being inundated with mission requirements. Sometimes it gets pushed to the side. • The tempo is high in this base, and some of the squadrons are finding it difficult to participate because of constant missions.
Location	Installation characteristics: • The military turnover is high here, and the civilian personnel have been here a really long time . . . and they don't seem to be interested in changes. • FSS facilities' employees don't want to come outside the box and make accommodations for the squadrons. Outdoor Rec has only 50 paintball guns and had 200 people in one event. You can just do it in sections, but [Outdoor Rec] wanted to do an event for only 50 people. It is exhausting. Local area: • When you are in a remote location, it is a lot more difficult to spend the money than if I was in Vegas or Florida or a highly populated place. We don't have a place for people to go and do everything. Most of what we have are portable options, and we have to find a place to put that portable option. Or the available places are more like a kids' place. Adult recreation activity options are limited, so you need to be creative. • The challenges you have OCONUS are totally different from the stateside challenges. We [OCONUS locations] have currency exchange issues that cut into the funding . . . The security issues are different as well. The C3 in [COUNTRY] has to do everything on base.

SOURCE: Interviews with C3s in 2019.

Summary of Findings from C3 Interviews

Interviews with C3s provided insight into the nature of C3 roles and the process of planning a UNITE event. Although certain aspects of the event planning process are fairly consistent across installations, these interviews highlighted the ways that C3s have tailored their procedures according to the characteristics of their installation (e.g., MAJCOM, size, availability of resources in the community) and the preferences of the units they work with. The perceived challenges provide some insight into potential changes that could be made to the initiative and/or resources available to ensure that UNITE events are implemented successfully. These changes include increased availability of training materials and/or forms that C3s could use to communicate about UNITE with POCs at their installations, increased marketing support, and streamlined procedures for obtaining AAR and participation information from POCs.

Community Cohesion Coordinator Email Survey

In addition to the in-depth C3 interviews described in this chapter, all C3s received a short email survey roughly three to six months after the interviews were conducted. The survey consisted of six open-ended questions:

- How often do commanders come to you with a specific UNITE activity in mind?
- How often do you give commanders suggestions for UNITE activities?
- When you give suggestions to commanders, how do you identify potential activities?
- What is the single biggest obstacle you face as a C3?
- What single thing would you change about UNITE to make it more effective or efficient?
- Is there anything else you would like to share about UNITE?

The C3 interviews, because of their in-depth nature, were the primary source of data for understanding the implementation of UNITE and the process of executing a UNITE event. However, the email survey data were collected in February and March 2020, providing us with more-recent data about the functioning of UNITE, relative to the in-depth interviews with C3s. The survey items were designed to solicit more-targeted information about the C3-commander interaction when planning an event and to gain additional perspectives on implementation challenges. At the time, there were 73 C3s; of those, we received responses from 40, for an overall response rate of 55 percent.[11] Responses to each item were relatively brief (i.e., one to three sentences).

The results discussed in the following section do not include raw numbers or percentages of C3s who endorsed specific ideas; however, example quotes are provided. Although the response rate was exceptionally high for a survey, we cannot rule out a bias associated with who did and did not respond. If there is a systematic bias in who responded, the example quotes presented in the next section would not represent the universe of all C3s.

Results

Here, we address each of the survey questions and provide example quotes when appropriate.

How Often Do Commanders Come to You with a Specific UNITE Activity in Mind?

C3s interpreted this question in one of two ways. First, some C3s provided information about whether a unit commander or a designated POC came to them. Comments indicated that POCs were more likely to engage with C3s than commanders were. Second, other C3s provided information about whether units (either commanders or POCs) came to them with specific ideas for UNITE activities. The range found in the open-ended responses here was quite broad, from never to "nearly 100% of the time."

[11] Of the 40 C3s who completed a survey, 17 also had participated in an interview.

How Often Do You Give Commanders Suggestions for UNITE Activities?

C3s also interpreted this question in one of two ways. First, as with the prior survey question, some C3s provided information about whether they interacted with a unit commander or a designated POC. Although this interpretation was less frequent than with the previous question, the result was the same: POCs are contacting C3s more than commanders are. Second, C3s provided information about how frequently units (either commanders or POCs) were given suggestions for UNITE activities. Coding of open-ended responses also indicated a range, from almost always, constantly, or every time to a very specific low of 20 percent.

In terms of mechanisms through which C3s provide examples of acceptable activities, C3s noted that they do so when they brief commanders or POCs, put examples on UNITE installation websites, and provide examples in newsletters. Comments reflected the importance of briefing new commanders as they transition to a new installation and the importance of continuously briefing commanders and POCs (e.g., described as "an ongoing process" by one C3).

When You Give Suggestions to Commanders, How Do You Identify Potential Activities?

We identified four themes across responses to this survey item. First, C3s discussed recommending on-base versus off-base activities:

- When giving activity suggestions (again, typically with POCs), I always begin with promoting on-base FSS options, such as golf, bowling, etc. I then will go into local, off-base options that I am familiar with or that have been approved in the past.
- I attach two documents. One that lists all the activities we offer on base, while the other lists off-base activities.
- First, I'll ask if they want to stay on base or go off base. Based on their answer, I'll provide my list of ready-to-go options. If they're looking for something else, we'll sit and brainstorm together.

Second, C3s noted that they recommend activities according to the distinct characteristics of the unit, including the size of the group, work environment, and work schedule:

- I go off the size of the head count. If it's a complete squadron, I program for bigger venues (i.e., squadron fun day), [but for] smaller work center type events, I use the bowling alley, Club or smaller ODR [Outdoor Recreation] equipment rental.
- I normally take into account their work schedule and the size of the event that they would like to do, i.e., is it just a shop/office, or the squadron as a whole, since some facilities have constraints on numbers and time to get everyone through. Once I know those factors, it makes it a lot easier to provide suggestions that are more tailored to that particular squadron.

C3s also described the specific needs of the unit planning a UNITE event:

> I typically ask what they need to work on in regards to resiliency and try to match that need up with an appropriate activity. If it's stress from the training tempo, I would suggest something relaxing, like a yoga and oils class. If it's communication, I think escape rooms are a fun way to work on that.

Third, C3s talked about providing units with preexisting lists of activities:

- I am very knowledgeable about activities that are available both on base and off base, so I have a large inventory of suggested ideas that was put together when the program was first rolled out. This is updated when new businesses are created.
- I have a compiled list of AFSVC [staff]-approved events (I keep it current to reflect latest updates/changes in [AFSVC] guidance and/or other governing directives/instructions/regulations). I have also created (since program start at our base) and made available a "menu" of activity ideas to include activities offered through our wonderful FSS facilities and the local community (includes things not offered on the base, like bowling, go karts, trampoline areas, hiking trails, etc.). All UNITE-related documents are posted on our FSS C3 UNITE SharePoint site for ease of access by the units/base personnel.

Finally, C3s discussed making sure that activity recommendations are linked to the core goals of UNITE (e.g., cohesion, team-building, camaraderie):

- Whenever I mention potential activities, whether it is to commanders or POCs, I try to enforce how the program will bring people together.
- Potential activities are ones that will promote unit cohesion, along with team-building and camaraderie. These include free, unit directed, already put together.

What Is the Single Biggest Obstacle You Face as a C3?

We identified seven themes in C3 responses to the survey item about obstacles that face C3s. Table 3.6 lists those themes with example quotes. The major themes include (1) UNITE funding amounts, (2) the AFSVC approval process for UNITE events, (3) inconsistent policy and guidance from AFSVC, (4) unit-specific issues (e.g., last-minute request, wanting to do something not aligned with UNITE goals, changes in scheduling caused by a mission), (5) POC-specific issues (e.g., inconsistency in POCs, assignment of POCs), (6) awareness of UNITE (e.g., marketing, new initiative, credibility with command), and (7) manpower (e.g., expanded duties, such as planning non-UNITE events on an installation).

Table 3.6. Email Survey: What Is the Biggest Obstacle You Face as a C3?

Identified Obstacle	Example Quote(s)
Funding	• The $13.50/$5 fund limit per person. Many units have said that they should be able to use their funds how they see fit. Still staying within the rules of the types of activities you can do, but not being limited to the low per person amount. Example: If a unit wants to do an activity that is $20 per person, they are allowed to make that call.
Approval process for events	• The single obstacle was getting approval for an event approved through AFSVA [AFSVC]. The process is not efficient. • With the new CY20 [calendar year 2020] Program changes, the biggest challenge I think will be the 21 day prior to event submission for Event Proposal Forms.
Inconsistent policy or guidance from AFSVC	• No standardization from [personnel at] AFSVC with regards to money or types of events we can do. Their DCSs [Defense Collaboration Services] are confusing, and we have to communicate via Facebook, which isn't the best method. • The rules keep changing . . . It makes it very hard to give a singular message when you have to go back each time and tell your audience that the rules have changed yet again.
Unit-specific issues	• Squadrons playing by the rules and ensuring that their activities are meeting the intent of the program. • The biggest obstacle is working with units and appointed POCs that wait until the last minute to plan and submit events. • POCs/Commanders finding time to participate in UNITE events. Everyone seems excited to have the money to use at first, but they have a hard time committing to a day or time because of the mission requirements.
POC-specific issues	• Not having dedicated POCs, every event I have to work with a new POC for the squadron and, with the realignment, most of the commanders have changed, this is an ongoing process. • Not having one POC for each squadron that all request[s] are funneled through. When squadron commanders allow for the different offices to split up the UNITE events, you get a new POC for each event, and most times no information is shared with them and that causes a lot of backtracking. • Getting the Squadrons to appoint POCs. That seems to be my biggest obstacle. Once they do and we plan their event and they see how easy it is, they come back.
Awareness of UNITE	• Communication with the squadrons. I need to get out more to meet with the commanders, Chiefs, and First Shirts. • The biggest obstacle I face is making sure the word is out about UNITE at my base. It is such a large base and can be tough to reach everyone I need to reach. • The lack of knowledge at the command level of the program. It is still a struggle getting squadrons to become invested in the program.
Manpower	• Not having a team or an assistant—GS5 or below would be GREAT!!!! It would be great if UNITE was its own flight . . . • The additional duties, or having to plan base events on top of running the UNITE program. It gets a little overwhelming, especially during the end of the year when all the requests roll in to balance both, when other C3s are solely responsible for just the UNITE program. • Time—supporting other programs detracts from time devoted to UNITE.

SOURCE: 2020 email survey of C3s.

What Single Thing Would You Change About UNITE to Make It More Effective or Efficient?

The survey asked C3s what specific changes they would make to UNITE to make it more effective or efficient. Their responses clustered around six themes: (1) funding, (2) data collection, (3) AFSVC (e.g., approval process, UNITE command structure), (4) local authority, (5) standardization (e.g., of UNITE materials and processes), and (6) information-sharing (e.g., marketing and awareness). Table 3.7 provides example quotes for each theme.

Table 3.7. Email Survey: What Single Thing Would You Change About UNITE to Make It More Effective or Efficient?

Identified Change	Example Quote(s)
Funding	• Provide more APF money per person. $13.50 isn't enough to do much of anything off base and it'll not really force people to use FSS activities. • More effective = more money. It is difficult to find programs that people are interested in doing for under $20 per person and airmen . . . will [not want] to use booster club funds or personal funds to pay for "mandatory fun."
Data collection	• Also, the websites that we use for submission are horrible. We need something that is specifically for UNITE and tracks items properly and has the bandwidth to accomplish the task. • The way events are inputted. The program needs a dedicated IT [information technology] person. Putting events into two different areas is not efficient. The SharePoint site did the trick and needs to come back with its own IT support.
AFSVC	• I understand the daunting task of reviewing and approving (or denying) event proposals from bases all over the world, but participation and implementation would be better if there was a quicker turn on funding requests. • One POC for all event[s] in each squadron and maybe approval turnaround from the 21-day window that started in 2020. I would say 10 days max would be much more beneficial at the base level. • Expecting POCs and Commanders to plan 21 days in advance is kind of ridiculous considering mission requirements. If they decide on Tuesday they can get away for a UNITE event Friday, they should be able to execute these funds.
Local authority	• Allow C3s locally to have funding and events that have been approved in the past not need to be approved again by [personnel at] AFSVC. Let the bases truly own the program like we do for small morale funds that the squadrons can get. • To allow the individual C3s to approve/decline event requests. This may be difficult as not all C3s see an event the same. An event that the Program Coordinators may disapprove, a C3 at a particular base may think that it would be the perfect unit cohesive event. • It would be great if bases could come up with a number of specific programs that we can approve at our level and are only tracked at Air Force level. This would free up Services Center resources so they can provide more oversight over new programs.

Identified Change	Example Quote(s)
Standardization	• Standardized documents that are modified at the base level. We should have been given standardized request forms, AAR forms, funds tracking forms, etc. Instead, C3s created their own, and they shared them with the rest of the community. • I would have a training video that is created from AFSVC/HQ and sent to the field. This video should highlight the program intent, mission, and execution plans. This training video could be used by all C3s to train appointed POCs. Because leadership created the video (versus C3s at their individual installations), the messaging would be clear and consistent.
Information-sharing	• Share information consistently and professionally to all through official email. Individual C3s have asked questions that are relevant to the group through various means. That information is not being received by all, so everyone is not on the same playing field but on their own interpretations and game plans. • Getting the word out Air Force–wide since everyone is now included. • More of an AF [Air Force] push to market this program.

SOURCE: 2020 email survey of C3s.

Is There Anything Else You Would Like to Share About UNITE?

C3 responses to the final survey question clustered into three types: positive feedback, negative feedback, and recommendations. One C3 provided a specific example of the positive aspect of UNITE:

> It is a very viable program and appreciated by those using it as intended. To have a POC remark that the airmen he works with are stressed because of the high mission tempo and additional hours being work and many with suicide ideation, and that he is grateful there is some funding to show there are those that care—hits it home that we can make a difference with someone's life.

Although C3s were largely positive in their comments, others pointed out negative aspects. Many of these comments were about procedural aspects of UNITE and have already been described here (e.g., the approval process, lack of standardization, lack of flexibility at the installation level). However, one comment pointed out a more fundamental problem:

> While I believe the UNITE program is a great idea, but I do not believe we are reaching its potential as a true resiliency or moral[e]/welfare program. If that's not the intention of UNITE, then don't say that's the intent of the program in training or use CSAF [Chief of Staff of the Air Force] quotes as revitalization of Sq's [squadrons].

Although this is one example, other data from commanders and unit POCs (i.e., via AARs) suggest that the intended purpose of UNITE may not be reaching airmen, as we will discuss later in this chapter.

Finally, some C3s offered specific recommendations for UNITE. We offer three of those recommendations here. The following examples were selected because they could lead to actions that could be taken immediately in an attempt to improve the efficiency and effectiveness of UNITE:

- Maybe approve funding for travel/mileage expenses for C3 programmers scouting and observing programs (we don't have access to a government vehicle).
- It would also be nice to get the personnel breakdown numbers used to allocate funds to the base, per squadron, from AFPC [Air Force Personnel Center] to streamline the new-year funding process.
- More emphasis from Wg [Wing] Commanders to use them [UNITE funds]. Commanders and POCs are having a hard time planning when to use the funds because they don't feel comfortable asking to close down. This has led to the funds getting used in small groups as opposed to squadrons, and I don't believe that was the intent of the program.

Summary of Findings from Email Survey

The email survey of all C3s provided supplementary data to the more in-depth interviews completed earlier in the study. These surveys showed us that C3s largely interact with POCs, not unit commanders, though many units already have specific activities in mind when they approach C3s. When C3s offer advice, many share preexisting lists of potential activities (often activities that other units have completed) but acknowledge that activities should meet each unit's needs. C3s told us that they offer both on- and off-base options but that they make sure that those options align with the core goals of UNITE: cohesion, team-building, and camaraderie. C3s reported several issues with UNITE, including those related to limited funding for food, timeliness of activity requests and approvals, inconsistent policy and guidance, lack of awareness of UNITE, overburdening of C3s, and assignment of POCs within units, many of which were also raised in the interviews. Not surprisingly, these are the same areas in which C3s suggested that things need to change to make UNITE more effective and more efficient.

After Action Reports

After each UNITE event, C3s requested responses to a set of five questions from each unit POC or commander who was involved with the event, as follows:

- What FSS or off-site establishment(s) did you partner with for this event?
- What went well for this event?
- What areas needed improvements for this event?
- Would you do this event again? Why or why not?
- What lessons were learned, and what recommendations do you have for future squadron events?

Although the AARs are not collected for evaluation purposes, they provide an opportunity to include insights from POCs and commanders, from whom we were unable to collect primary data. To explore these data, we analyzed open-ended responses to the questions about what went well, what improvements are needed, what lessons were learned, and recommendations for future events taken from a random sample of the 2,452 events that took place during our analytic period.[12] We did not examine the list of specific vendors, nor did we examine data from the question about repeating events. A preliminary review of the data revealed that very few respondents (i.e., less than one-half of a percent) indicated that they would not do the event again. The preliminary review also suggested that there could be useful information in response to the "why" part of the question but that this information likely was largely redundant with other questions in the AARs. Therefore, we chose not to review these data in detail.[13]

Before reviewing the results of the AARs, we highlight four possible limitations associated with this data source. First, as noted, these data were not collected for evaluation purposes. Therefore, comments were generally brief and designed to provide feedback to AFSVC staff rather than to provide formal feedback on the process of planning and executing an event (unlike the C3 interviews). Second, some comments may be impressions provided by C3s rather than direct input by POCs or unit commanders.[14] What went well, what did not go well, and recommendations for the future of UNITE may vary greatly between these two groups, given their different vantage points of the unit. Third, some events were combined with other unit-wide events, such as a resilience day or a stand-down because of a suicide. Thus, it is not entirely clear whether comments are really about the UNITE event, the other event, or both. Finally, the analysis used only a sample of all AARs. With more than 2,400 UNITE activities in our database that met our selection criteria, we could not code every open-ended response. Although our sampling strategy should provide us with a random sample of events, we did not capture every recommendation or lesson learned provided in the AARs. Detailed result tables can be found in Appendix B. Here, we briefly summarize the main themes from the AAR data.

Summary of Findings from After Action Reports

Overall, units' comments in the AARs suggested that they were satisfied with many aspects of UNITE. The execution of planned events, having time to socialize and interact with other unit members, and having an opportunity to build unit morale and cohesion and engage in team-building activities all were specific aspects that respondents highlighted in their comments about what went well.

Although units expressed a great deal of satisfaction with UNITE, they also identified some areas for improvement. Misunderstandings about how to use funding and a desire for more flexibility in what UNITE funding can be used for were reported by some respondents. Others identified issues with participation, including maintaining an accurate head count ahead of time

---

[12] We used a random number generator to select up to four events from each installation that hosted UNITE events during the analytic period. For more details, see Appendix B.

[13] The AARs included other data, such as brief event descriptions; however, the limited details provided in these event descriptions prevented us from using these data to categorize the types of events that took place during the study period.

[14] In this section, we simply refer to "units" rather than commanders or POCs, since we do not know who provided the information for the AAR.

for planning purposes, low attendance, and scheduling conflicts. One of the more ubiquitous themes in comments for this AAR question was the need to plan further in advance. Some units noted that this early planning process would allow time to consider external factors that might influence attendance and participant satisfaction (e.g., weather, venue kitchen size) and to develop a marketing plan to increase interest and participation by unit members. Finally, some units noted the importance of making sure that the reason for UNITE events is made clear to attendees during events.

AAR comments offered several lessons learned and recommendations for future UNITE activities. Although these comments often were event-specific (e.g., "Restaurant X was great and everyone should use them!"), others were more generalizable to UNITE as an initiative. Hosting events outside the work environment, communicating frequently with a host venue, selecting activities that have a wide appeal to all unit members, planning well in advance of an event, having a backup plan in case of unexpected events, using marketing and scheduling to increase attendance, understanding the nuances of UNITE funding, and making sure UNITE events have a purpose that is well-known to attendees all were recommendations made by units to make UNITE a more effective and more efficient initiative.

Open-Ended Participant Survey Data

Approximately two weeks after participation in a UNITE activity, airmen who participated in the activity and who provided a DoD ID that could be matched to a DoD email received an invitation to take a short, two-minute survey (see Appendix C for a full description of the survey).[15] At the end of the survey, respondents were offered an opportunity to provide any additional thoughts they had about UNITE in an open-ended text field.[16] This text field was not designed as a formal survey item but was required by the Air Force Survey Office. Responding to the open-ended survey item was optional. We received 595 comments regarding activities that occurred between July 27, 2019, and November 7, 2019. Although this survey was not designed to be a formal data source for the process evaluation, these responses presented an opportunity to include airmen's qualitative impressions of UNITE in the evaluation (in addition to the quantitative data discussed in the next chapter).

After an initial review of the comments, we identified four items relevant to UNITE that we could address with the data:

- What is the impact of the program or activity?[17]
- What are the positive aspects of UNITE?

---

[15] Note that it is possible that a unit POC was included in the survey. For this analysis, he or she is considered a UNITE participant, not a POC.

[16] The item wording was, "Do you have any other general comments you would like to provide?"

[17] Note that findings from this question are reported in Chapter 4.

- What are the negative aspects of UNITE?
- Participant recommendations.

Not all survey respondents offered additional comments, and thus there is a risk of bias associated with this survey if respondents who offered a comment are systematically different from those who did not on some characteristic or UNITE-related experience.[18] This risk of bias is important to keep in mind when reviewing the results. The details of these results can be found in Appendix B. Here, we briefly summarize the main themes from these comments.

Summary of Findings from Participant Survey Data

According to participant comments, UNITE had a demonstrable perceived impact across several dimensions, including improved unit morale and camaraderie, increased unit cohesion and unity, and increased opportunities for team-building, socialization and interaction with other unit members, and relaxation. Participants were also generally positive about UNITE when they provided comments, expressing satisfaction with specific activities and vendors, expressing gratefulness for UNITE funding, and highlighting opportunities to interact and bond with fellow unit members while having fun and relaxing. Despite the general positivity toward UNITE, some participants offered a less enthusiastic take on the initiative. These comments pointed to problems with the logistical aspect of implementing an event, difficulties using and restrictions on funding for activities, unclear and mixed messages about the reasons for the initiative, and lack of awareness. Two of the more common complaints were related to participation: "forced" and low attendance. Finally, participant comments provided us with unsolicited recommendations for the future of UNITE. These included making funding easier to use and increasing the amount available (especially for food), allowing commanders and units more control and more flexibility over the activities they have, holding events during traditional duty hours to increase participation, holding events off the installation to fully relax and recharge, and increasing the frequency of UNITE events.

Satisfaction According to UNITE Post-Participation Surveys

Although the quantitative data from the post-participation surveys are the focus of the next chapter on our outcome evaluation of UNITE, one survey item is relevant to our process evaluation. Satisfaction is an output in the UNITE logic model. One item on the survey asked airmen to rate the following sentence on a scale from 1 (strongly disagree) to 5 (strongly agree): "I was satisfied with the activity my unit participated in." The average score among survey

---

[18] Respondents who provided substantive comments were older, more likely to be officers, and had served longer than respondents who did not provide comments. However, the two groups did not differ with respect to ratings of cohesion.

respondents was 4.53 (standard deviation of 0.83), suggesting that airmen were satisfied overall with the UNITE event they attended.

Summary

In this chapter, we reviewed qualitative data gathered from C3s, units, and UNITE participants to assess how UNITE is being implemented. Using the logic model presented in Chapter 3, we can compare what we heard from these various sources with the resources and inputs, activities, and outputs that UNITE is expected to use and generate. Shortcomings in these areas may lead to recommendations that can improve the efficiency and effectiveness of UNITE.

Table 3.8 summarizes the key takeaways from the C3 interview, C3 email survey, AAR, and open-ended participant survey data. We can map these findings onto the process elements of the UNITE logic model.

Resources and Inputs

The logic model identifies six key resources and inputs used by UNITE:

- C3s
- unit commanders and POCs
- C3 training, manuals, and guides
- UNITE CONOPS
- funding
- an event tracking system.

In terms of personnel, C3s were largely viewed as helpful by unit leadership and UNITE participants. However, we heard from some C3s who are overburdened, sometimes being tasked by installation leadership with duties outside of UNITE. Commanders and unit POCs play an important role in the UNITE planning process and in providing data needed for evaluation, though it was not always clear who was providing that data and how involved unit leadership is. Turnover of unit leadership is also a potential concern, and C3s noted that they must work continuously to maintain awareness of UNITE among leadership.

C3s were largely complimentary of the training they received, though we heard some complaints about a lack of subsequent interaction with other C3s after training, especially via existing networking tools (e.g., Blackboard). The one area where we heard consistent criticism was the inconsistent and ever-changing guidance provided by AFSVC and, in particular, what was and was not covered by UNITE funding. Although many participants were grateful for the funding, all the various types of informants noted that increasing the amount that could be used on food could increase both satisfaction and participation.

As we will discuss in more detail in Chapter 5, the current event tracking system may be in need of an update now that the initiative has been in place for some time. Understanding what data is needed by various stakeholders (e.g., financial data, event data, outcome data), together with feedback from current system users, could inform these changes.

Activities

Four activities are part of the UNITE logic model: marketing of UNITE to unit leadership, recruiting commanders to use the initiative, identifying and coordinating events, and conducting pre- and post-event data collection in AARs. One consistent theme that we heard, across data types and informants, was that not enough is done to market UNITE as an option for commanders to use. Some C3s said they used their personal funds to create marketing materials, and others noted a lack of consistent marketing materials to use across the Air Force. As we discuss later, the awareness of UNITE may play a role in how many units use it. Messaging about the purpose of UNITE also was viewed as lacking by some C3s, commanders, POCs, and participants. Some of this messaging may need to address the notion that UNITE activities are just another form of "mandatory fun."

C3s spoke to how units identify events, telling us that they use a variety of methods to work with units in developing and planning activities. Although a one-size-fits-all approach may not be appropriate, the variability in the planning process could explain why UNITE is more (or less) successful at one installation versus another. One of the biggest complaints about the planning process was the lead time required to get a proposed UNITE event approved. This complaint led some C3s and unit commanders and POCs to suggest that perhaps not all events, such as FSS-based RTEs or others that are done frequently (e.g., bowling on the installation), need AFSVC approval. Many units would like more freedom over how UNITE funds are used.

Finally, data collection procedures are one area in which reality does not match the ideal. We heard frustrations with online collaboration tools and with the tools used to collect data about UNITE activities, both before and after they occur (e.g., AARs). C3s and units also complained about the collection of DoD ID numbers used to identify airmen who should receive the post-participation surveys. Together, these issues have real implications for program evaluation and, specifically, outcome evaluation.

Outputs

There are two outputs in the logic model: participation in UNITE (by both commanders and airmen) and satisfaction. As noted in the previous section, marketing is one area in which the reality of implementing UNITE may not match the ideal. A lack of awareness about the initiative could affect uptake on the part of commanders. Another consistent theme that is related to participation was frustration with a lack of attendance at UNITE events. Nonetheless, satisfaction with the initiative cut across data sources. Thus, although UNITE was clearly not perfect in the eyes of all C3s, commanders, and airmen, the majority viewed it positively. Post-participation survey data suggest that satisfaction with UNITE events was high.

In the next chapter, we focus on the rest of the UNITE logic model and review the results of the outcome evaluation, which relies primarily on airmen post-participation survey data.

Table 3.8. Key Takeaways from Qualitative Data

Resources and Inputs	Activities	Outputs
C3s: • C3s might have additional duties outside of UNITE but no additional staff to help. (C3E)	Marketing of UNITE: • There is a need for increased awareness of UNITE, and standardized marketing tools would be beneficial. (C3I) • Part of marketing is making sure that airmen understand why UNITE exists. (AAR) • Lack of awareness of UNITE may limit uptake. (C3E)	Participation in UNITE (by both commanders and airmen): • Low attendance can be a problem, but deconflicting schedules and hosting events outside the work environment and off the installation may help. (AAR) • Low participation was a concern, and participants offered off-site, duty-day activities as a way to increase it. (PS) • Funding for food at events was seen by some participants as too low and may have implications for attendance. (PS)
Unit commanders and POCs: • Engaging with units, especially as commanders and POCs change, is an ongoing process. (C3E)	Recruiting commanders to use the initiative: • C3s vary in their approach to communicating with commanders and POCs and planning events. (C3I)	Satisfaction: • Units reported that UNITE activities were a time to socialize and interact with other unit members, build unit morale and cohesion, and engage in team-building. (AAR) • Overall, satisfaction was rated high. (PS) • Participants said UNITE activities led to opportunities for team-building, socialization and interaction with other unit members, and relaxation. (PS) • Some airmen still see UNITE activities as "forced" or "mandatory" fun and do not understand UNITE's purpose. (PS)
C3 training, manuals, and guides: • Initial C3 training provides a solid foundation for the role, though refresher training would be valuable. (C3I) • A lack of standardized materials and guidance leads to variation across installations (e.g., with respect to the planning process, preference for on versus off base). (C3I) • C3s report inconsistent and unclear policy and guidance from AFSVC. (C3E)	Identifying and coordinating events: • Features of the installation and local community shape the specific offerings. (C3I) • There is no one-size-fits-all approach to creating UNITE activities across units. (C3E) • The approval process for activities is cumbersome and not timely; last-minute requests by units cause frustration and may affect the quality of events. (C3E) • Units were frustrated with the amount of lead time required to hold an event, but those who started the planning process early reported a better experience. (AAR)	

Resources and Inputs	Activities	Outputs
UNITE CONOPS: • Lack of formalized guidance (e.g., an AFI) is a barrier. (C3I) Funding: • Amount and timing of funding can create logistical challenges (e.g., APF and NAF are on different calendars). (C3I) • The budget for food is viewed by some C3s as too low. (C3E) • Lack of flexibility in the use of UNITE funds is viewed by units as a limiting factor. (AAR) • Frustration with funding limitations led some participants to suggest that units should have more control over and flexibility in determining what activities can be done with UNITE funds. (PS) Event tracking system: • Data collection about events (e.g., AAR data from units, obtaining DoD IDs) is problematic in several ways, including ease of use problems, lack of information technology support, and nonresponsiveness on the part of data providers. (C3E)	Conducting pre- and post-event data collection in AARs: • Submission of AAR data is difficult because of the unwieldy data system and challenges getting the information from POCs. (C3I)	

NOTE: AAR = after action report data from unit commanders and POCs; C3E = C3 email survey; C3I = C3 interviews; PS = participant survey.

Chapter 4. Outcome: What Does UNITE Do?

In Chapter 1, we noted that the evaluation of UNITE stemmed from an interest in understanding how FSS experiences might foster unit cohesion, readiness, and resilience. To better understand the association between participation and unit cohesion, we used the evidence-informed framework described in Chapter 2 to develop two survey instruments for airmen to be completed immediately following participation in a UNITE event and a few weeks later (for details, see Appendix C). Specifically, our approach to designing a post-participation survey for the UNITE Initiative began with identifying which building blocks of readiness and resilience were most likely to be addressed in UNITE events. In addition, we were interested in whether the characteristics of the UNITE event matter for the building blocks and for cohesion.

This analysis represents an important first step to understanding whether UNITE is achieving its goal of increased unit cohesion via specific building blocks. Specifically, we were interested in exploring three research questions:

1. How are event characteristics directly associated with the readiness and resilience building blocks?
2. How are readiness and resilience building blocks directly and indirectly associated with overall unit, social, and task cohesion at both the first and second post-UNITE surveys?
3. How are event characteristics associated with overall unit, social, and task cohesion (both directly and indirectly) through the readiness and resilience building blocks?

To explore these questions, we analyzed quantitative data from airmen post-participation surveys. In addition, we examined open-ended data from the first post-participation survey to assess participants' thoughts on the perceived impact of UNITE.[19] In this chapter, we first provide a brief overview of the surveys, as well as the data and analytic approaches leveraged to answer the research questions. Then, we focus on the answers to those research questions.

Quantitative Analysis and Results

Data Sources

UNITE Post-Participation Survey

The Air Force administered two one-minute surveys to active-duty airmen after they participated in a UNITE event.[20] The first survey included 17 items and was administered within approximately two weeks of the UNITE event. The second survey included 11 items and was

---

[19] Detailed results can be found in Table B.1 in Appendix B.

[20] Full details on the development and administration of the UNITE surveys are provided in Appendix C.

administered approximately six weeks after the first survey (or eight weeks after the event). These surveys allowed us to examine whether the associations we examined changed over a brief period. Both surveys included the Group Environment Questionnaire (GEQ),[21] a validated measure of cohesion, including subdimensions of task and social cohesion. The first post-participation survey included single-item measures of six readiness and resilience building blocks that are targeted by UNITE to promote cohesion, as depicted in Table 4.1 (see also Appendix C): physical activity, involvement in activities, coping strategies and skills (specifically, decompression or leisure as coping), peer group and Air Force values, and social network (specifically, social interaction). The second post-participation survey also included an item about social interaction. Table 4.1 also shows how the building blocks map onto the short-term outcomes from the UNITE logic model and how they are operationalized in the post-participation surveys.

Table 4.1. Building Blocks, Outcomes, and Airman Survey Items

Readiness and Resilience Building Block(s)	Short-Term Outcome	Airman Post-Participation Survey Item
• Coping strategies and skills • Involvement in activities	• Positive use of leisure time	• Participating in this activity provided me with additional opportunities to interact or connect with members of my unit. (Survey 1) • Members of my unit are interacting and connecting more because of this activity. (Survey 2)
• Coping strategies and skills	• Opportunity to decompress	• This activity provided me with an opportunity to unwind (i.e., rest, relax, and/or have some fun).
• Peer group/unit values • Community/Air Force Values	• Promotion of Air Force institutional values	• Participating in this activity with unit members promoted or reinforced Air Force core values.
• Physical activity • Involvement in activities	• Increased physical activity	• Participating in this activity was physically demanding.
• Social network	• Increased social interaction	• Participating in this activity provided me with additional opportunities to interact or connect with members of my unit.

Personnel and Installation Data

The Air Force provided airmen demographic characteristics (e.g., gender, age) and military characteristics (e.g., rank) for each survey respondent from personnel records. Installation data

[21] See Carron, Widmeyer, and Brawley, 1985; Chang and Bordia, 2001; Gupta, Huang, and Niranjan, 2010; Treadwell et al., 2011; and Widmeyer, Brawley, and Carron, 1985.

included installation size (according to Air Force Personnel data) and remote or isolated status as defined in AFI 65-106.[22]

Event Data

UNITE event characteristics were gathered from the event descriptions that C3s submitted through the UNITE C3 SharePoint database (e.g., FSS versus non-FSS).[23] Each data source is described in more detail in Appendix D. A limitation of these event data is that they are relatively coarse (compared with personnel data) and do not always capture the nuances of events (e.g., detailed activity descriptions, specific amounts of funding). This issue is especially pertinent to our evaluation, given the variety of event types offered as part of UNITE. Further, interpreting these data proved to be more challenging than expected, hindering our ability to fully identify associations between event characteristics and cohesion. Therefore, we used a limited number of event characteristics that we expected to be more reliably reported from this source.

Measures

Airmen Demographic, Military, and Installation Characteristics

We used the Air Force personnel data to create indicators of airmen demographic characteristics, including gender and age. Similarly, we created indicators of rank. Installation characteristics include an indicator for remote or isolated installation and installation size as measured by the number of active-duty airmen at the installation.

Resilience and Readiness Building Blocks

The surveys also asked respondents to indicate the extent to which building blocks were related to the UNITE event in which they participated. Items used a five-point Likert scale that ranged from 1 (strongly disagree) to 5 (strongly agree), with a 3 indicating a neutral stance. In particular, the survey probed the following readiness and resilience building blocks that were posited to foster cohesion: increased physical activity, involvement in activities, coping skills and strategies (specifically, decompression or leisure as coping), social network (specifically, social interaction), and peer group/Air Force values. All of these items were fielded as part of the first survey. A question on social network (specifically, social interaction) was also fielded in the second survey.

---

[22] See Air Force Instruction 65-106, 2019.

[23] Note that we were unable to obtain unit-level information to include in our analysis. For example, we do not know how long a unit commander has been in place, nor do we know how long an airman has been assigned to their current unit (though we are able to assess whether an airman has moved in the year prior to participation in a UNITE event).

UNITE Event Characteristics

As noted in the previous section, the C3 SharePoint database provided information on UNITE event characteristics. Specifically, these files provided insight into whether the event was located on or off base, whether the unit leveraged MOA funds for the event, and whether the event was provided by the installation FSS.[24] The first post-event survey asked airmen whether their UNITE event occurred during duty hours. When working with units to design UNITE events, the unit commander or POC identified the reasons the unit was participating in UNITE from the following prepopulated list:

- to provide opportunity for fun or relaxation
- to promote interaction between unit members
- to increase morale, camaraderie, or esprit de corps
- to improve physical fitness
- to work on a team-building exercise
- to develop a new skill or competency
- to reinforce peer, unit/squadron, or Air Force values.

These reasons also map onto several of the resilience and readiness building blocks (see Table 4.2) but from the unit's perspective rather than the airman's perspective.

Table 4.2. UNITE Event Purpose and Building Blocks Crosswalk

Unit Commander UNITE Event Purpose	Readiness and Resilience Building Block
• To provide opportunity for fun or relaxation	• Coping strategies and skills
• To promote interaction between unit members	• Social network • Sense of community
• To increase morale, camaraderie, or esprit de corps	• Sense of belonging • Sense of community • Social capital • Social support • Community/Air Force values
• To improve physical fitness	• Physical health
• To work on a team-building exercise	• Social capital • Social support • Sense of community
• To develop a new skill or competency	• Coping strategies and skills
• To reinforce peer, unit/squadron, or Air Force values	• Peer group/unit values • Community/Air Force values

[24] MOA funds refer to APF that was provided via MOA for the implementation of UNITE.

Overall Unit Cohesion

As stated earlier, our primary outcomes were the GEQ scales, which provide a measure of overall unit cohesion. Airmen responded to survey items that corresponded to task and social cohesion on the first and second surveys to obtain a measure of overall unit cohesion shortly after UNITE participation (within two weeks) and six weeks after the first survey (within eight weeks of the UNITE event). Cohesion was measured on a five-point Likert scale, from strongly disagree (1) to strongly agree (5), with a 3 indicating a neutral stance. The overall unit cohesion measure is an average of the social and task cohesion subscales.

Analytic Sample

Our main analyses included a sample of 624 airmen who responded to both the first and second post-event surveys. When creating this sample, we excluded survey responses from airmen who could not be matched to a specific UNITE event in the C3 SharePoint database (n = 56). This matching was done by linking the event date and name, the installation where the event occurred, and the respondent's Air Force unit in both the survey and C3 database. We present additional data on a larger sample of airmen, which includes all participants who responded to the first post-event survey, in Appendix D. According to the participant survey data, 545 events are included in the first post-event survey, and 338 events are covered by both the first and second surveys.

Descriptive Statistics

Table 4.3 provides descriptive statistics for all of the measures and the sample described in this chapter. Approximately 76 percent of survey respondents were male and were, on average, 32 years old. Noncommissioned Officers (E5–E6) were the most common ranks (29 percent) in the sample, and Senior Airman (E4) was the least common rank (7 percent). Nearly one-quarter (22 percent) of respondents were assigned to a remote or isolated installation, and most respondents were assigned to a medium-sized installation (70 percent). According to the survey, airmen rated UNITE events as most associated with decompression, social interaction, and peer/Air Force values and, to a lesser extent, with physical activity. These outcomes align closely with commanders' intent for the activity. More than 90 percent of commanders reported that the events were meant to provide an opportunity for fun or relaxation, promote interactions among unit members, and increase morale, camaraderie, or esprit de corps.[25] More than half of commanders reported that the events were meant to work on team-building or promote peer/Air Force values. Only about one-third of commanders reported that the events were meant to promote physical activity or gain a new skill or competency. A majority of respondents participated in events during duty hours (85 percent), about half of respondents participated in

---

[25] Because of the lack of variation in the commander intent variables and their overlap with the building blocks, we did not use these variables in the final models.

events that were located off base, about two-thirds of participants participated in events that used MOA funds, and about half of the events were provided by the installation FSS. Finally, respondents reported that cohesion at both time points was, on average, between neutral (3) to positive (4) on the five-point Likert scale.

Table 4.3. Descriptive Characteristics of Airmen in the Analytic Samples

	First and Second Survey Respondents (*N* = 624)	
Airmen Demographics and Military Characteristics	**Mean[a]**	**SD**
Male	76.1%	--
Age	32.189	7.669
Rank (pay grade)		
Airman (E1–E3)	15.5%	--
Senior Airman (E4)	7.1%	--
Noncommissioned Officer (E5–E6)	29.0%	--
Senior Noncommissioned Officer (E7–E9)	20.2%	--
Company Grade Officer (O1–O3)	13.3%	--
Field Grade Officer (O4–O6)	14.9%	--
Installation Characteristics		
Remote/isolated	22.3%	--
Size		
Small (up to 1,000 active-duty airmen)	6.7%	--
Medium (1,001–5,000 active-duty airmen)	69.7%	--
Large (5,001–6,000 active-duty airmen)	12.3%	--
Mega-large (6,001–30,000 active-duty airmen)	11.2%	--
Building Blocks		
Physical activity	2.497	1.16
Coping skills and strategy: involvement in activities (decompress)	4.413	0.893
Positive use of leisure time (off-duty events only)	4.480	0.955
Social network (social interaction), Survey 1	4.527	0.787
Social network (social interaction), Survey 2	3.923	0.968
Peer/Air Force values	4.088	0.96
UNITE Event Characteristics		
During duty hours	85.3%	--
Off base	47.4%	--
Used MOA funding	68.1%	--
FSS-provided event	51.9%	--
Commander reasons for UNITE event		
To provide opportunity for fun or relaxation	92.6%	--

	First and Second Survey Respondents (*N* = 624)	
To promote interaction between unit members	94.5%	--
To increase morale, camaraderie, or esprit de corps	97.6%	--
To improve physical fitness	36.6%	--
To work on a team-building exercise	57.8%	--
To develop a new skill or competency	31.1%	--
To reinforce peer, unit/squadron, or Air Force values	57.9%	--
Unit Cohesion		
Social cohesion, Survey 1	3.281	0.738
Task cohesion, Survey 1	4.055	0.748
Overall unit cohesion, Survey 1	3.668	0.638
Social cohesion, Survey 2	3.338	0.762
Task cohesion, Survey 2	4.123	0.751
Overall unit cohesion, Survey 2	3.731	0.657

SOURCE: RAND analysis of airmen survey data.
NOTE: SD = standard deviation. Survey 1 refers to post-event surveys fielded within two weeks of UNITE event participation. Survey 2 refers to post-event surveys fielded eight weeks after UNITE participation.
[a] Categorical variables are presented as percentages without accompanying standard deviations.

Analytic Approach

An ideal evaluation of UNITE would include analyses that compare the cohesion of airmen in units that participated in UNITE with the cohesion of airmen in similar units that did not participate. A second-best approach to evaluating UNITE would be to measure cohesion shortly before UNITE participation (e.g., two weeks prior) and then again after UNITE participation (e.g., two and eight weeks after). Unfortunately, these approaches were not feasible, given the timing of UNITE implementation and the data collection approaches identified as achievable by the Air Force (i.e., only surveying UNITE participants). Therefore, our analyses focused on post-event participation ratings of cohesion and the building blocks of readiness and resilience for only the individuals who participated in a UNITE event.

We employed the statistical technique of path analysis to answer our three research questions. Path analysis, a type of multiple regression analysis, models the direct and indirect associations that exist among a set of variables. In our case, the path model quantifies the associations among UNITE event characteristics, building blocks, and cohesion.

We present a simplified version of our analytic model in Figure 4.1, in which each arrow represents an association that is estimated in our analysis. For example, arrow (or path) A allows event characteristics to be directly associated with cohesion as measured on the second post-event survey, arrow B allows event characteristics to be directly associated with building blocks, and arrow C allows event characteristics to be directly associated with cohesion as measured on the first post-event survey. Such paths are called *direct associations*.

Path analysis also allows for the estimation of indirect associations, or how one variable is associated to another *through* one or more other variables. In Figure 4.1, we posit that event characteristics should have a direct association with the select group of building blocks assessed in the survey (path B). The building blocks should then have a direct association with cohesion as measured at the first survey (path E). However, some of path E may be caused by the fact that event characteristics are also influencing the building blocks. So, the indirect association between event characteristics and cohesion at the first survey is some combination of paths B and C. That is, event characteristics operate on cohesion through the building blocks.[26]

Finally, the total associations are an estimate of the overall association, taking all possible paths (both direct and indirect) into account. To continue our example, the total association between event characteristics and cohesion from the first post-event survey would be a combination of paths B, C, and E.

The difference between the model depicted in Figure 4.1 and our full analytic model is that Figure 4.1 shows only one path between the event characteristics and the building blocks and only one path from the building blocks to cohesion. In our analytic model, each event characteristic has its own arrow connecting to each building block and both cohesion measures. Each building block also has its own set of arrows connecting to cohesion from the first survey to cohesion from the second survey.

Figure 4.1. Simplified Path Analysis Model

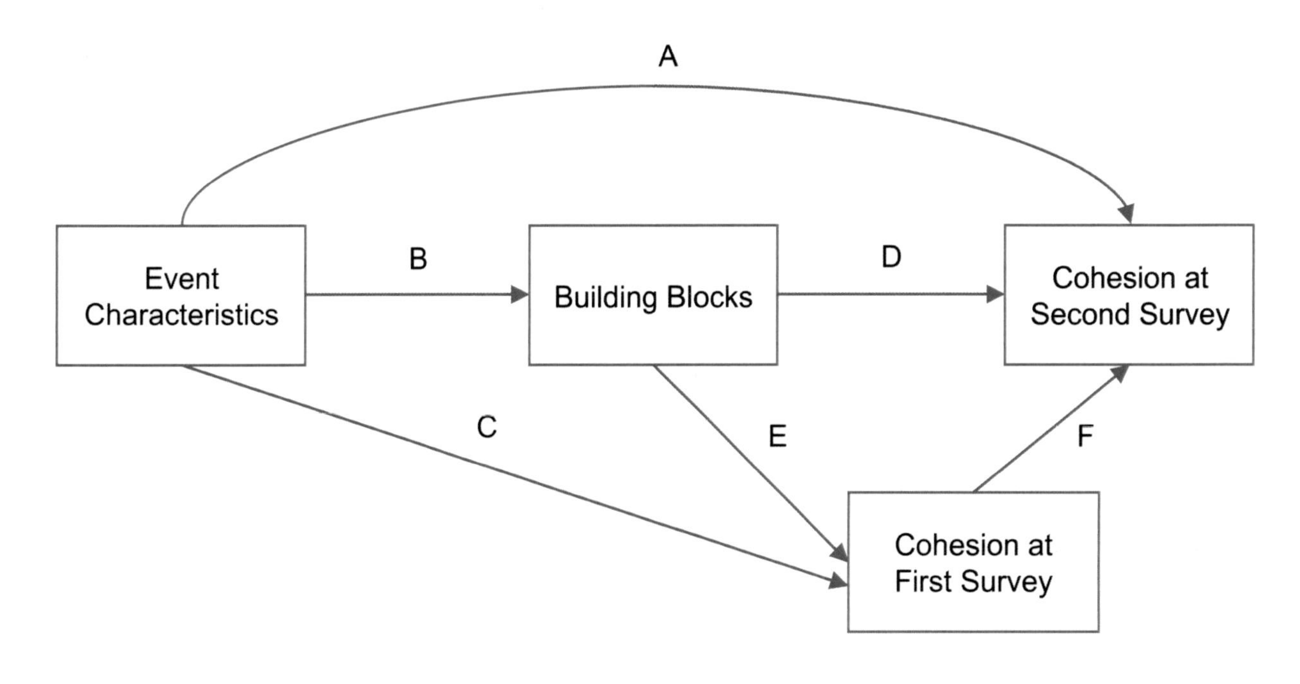

---

[26] As a more concrete example, one could hypothesize that events that are held off base are viewed by participants as more relaxing, leading to higher ratings of cohesion post-event.

Although the model in Figure 4.1 is a simplified version of our complete analytic model, we use it to provide a basic overview of how we leverage the various paths to answer each of the three research questions. Table 4.4 provides each of the research questions and the corresponding paths from the simplified model in Figure 4.1.

Table 4.4. Research Questions and Corresponding Paths in Path Analysis

Research Question	Corresponding Paths from Figure 4.1
1. How are event characteristics associated with the readiness and resilience building blocks?	• Path B
2. How are readiness and resilience building blocks directly and indirectly associated with overall unit, social, and task cohesion at both the first and second post-event surveys?	• First survey: path E • Second survey: path D • Total: paths D, E, and F
3. How are event characteristics associated with overall unit, social, and task cohesion (both directly and indirectly) through the readiness and resilience building blocks?	Cohesion first survey: • Direct: path C • Indirect: paths B and E • Total: paths B, C, and E Cohesion second survey: • Direct: path A • Indirect: paths B, C, D, E, and F • Total: All paths

In the main body of this report, we present results on overall unit cohesion and note if results differ for task or social cohesion. For simplicity, we also only show paths with associations that were statistically significant. Full results on all cohesion outcomes and results that include the full sample of airmen who responded to the first post-event survey can be found in Appendix D. Finally, all estimates are presented in effect sizes. An *effect size* is a measure of the size of the association that a variable (e.g., a building block) has on a desired outcome (e.g., cohesion), expressed as the fraction of a standard deviation of the outcome. A *standard deviation* is a measure of how much the outcome varies, or how spread out the outcome data are from the average. Thus, a larger effect size, or larger fraction of that standard deviation, implies a stronger association. Although a disadvantage of effect sizes is that they are difficult to interpret on their own, the advantage is that they allow us to compare results across studies, which often use different measures, on a common scale.

Limitations

Because both surveys were voluntary, response bias was a potential threat to our ability to interpret results from the analysis as related to UNITE versus some characteristic(s) of airmen. In other words, if older airmen were more likely than younger ones to participate in the survey and they also happen to be more engaged with their units, whatever association we find between UNITE and cohesion outcomes may not be caused by participation in UNITE events. We guard against this potential bias by controlling for key individual, military, and installation characteristics in our models. Those characteristics are gender, age, rank, remote or isolated installation status, and installation size.

Another key limitation is that not all airmen who participated in an event during the time frame of interest received a survey invitation.[27] To field this survey, the Air Force Survey Office used DoD ID numbers that were provided by UNITE participants and then submitted by C3s through the AAR. There were two key issues related to these data. First, not all participants were willing to provide their DoD ID to their unit POC and/or installation C3. Unfortunately, we were unable to identify the number of UNITE participants who opted not to provide a DoD ID, because the data we had available did not enable us to identify the exact number of active-duty personnel who participated in each event. There was also some data loss caused by inaccurate DoD IDs (e.g., DoD IDs were provided on a handwritten list, and some IDs may have been difficult to decipher). Thus, when AFPC received the list of DoD IDs, some of the IDs could not be matched to personnel records. (More detail on these limitations is provided in Appendix C.)

Although these controls account for pertinent factors that may influence estimated associations between building blocks or event characteristics and cohesion, they are not comprehensive. There may be other characteristics of airmen that our model does not account for that may influence our results. There may also be unaccounted-for unit, installation, or UNITE event characteristics that would influence airmen perceptions of cohesion. For example, units that are already high on cohesion may be more likely to participate in UNITE. Therefore, we cannot say that the building blocks or event characteristics *affected* cohesion. Rather our results indicate whether there is an *association* among those variables. Additional research is needed to determine the causal relationship between building blocks or event characteristics and cohesion.

Path Model Results

We ran one analytic model and used aspects of that model to answer each of the three research questions. Notably, when building the model, we removed event characteristics that were never significantly associated with building blocks or cohesion from our model for parsimony. These excluded variables include whether the UNITE event occurred during duty hours, whether it was held off base (event characteristics), and whether the event involved physical activity (building block). All models also control for gender, age, rank, being assigned to a remote or isolated installation, and installation size (according to the number of active-duty airmen). Results for each research question are presented in this section. Positive use of leisure time (building block) was also excluded from the final model because it was only asked of airmen who participated in UNITE events held outside of duty hours, which resulted in a small number of cases ($n = 92$).

---

[27] A related limitation is that some of the events that occurred during the period that our analysis covered do not have associated survey data from any participants.

How Are Event Characteristics Directly Associated with the Readiness and Resilience Building Blocks?

Research question 1 asks how event characteristics are associated with the readiness and resilience building blocks. Figure 4.2 presents the results of our analytic model, highlighting the paths that are relevant to this research question. Our results suggest that UNITE events for which MOA funds were used were positively associated with the decompression building block. The estimate of 0.261 ($p < 0.05$) indicates that airmen rated events that used MOA funds higher on decompression compared with those that did not use MOA funds, controlling for all other variables in the model. Using MOA funds was not significantly associated with the social interaction or peer/Air Force values building blocks. The indicator for whether the event was FSS-provided was negatively associated with a larger number of building blocks. Events that were FSS-provided were rated *lower* on the decompression, social interaction, and peer/Air Force values building blocks by 0.207, 0.181, and 0.186 standard deviations, respectively ($p < 0.05$), compared with events that were not FSS-provided. We show that these results are identical for task and social cohesion in Appendix D.

Figure 4.2. Path Analysis Results Related to Research Question 1, Overall Unit Cohesion

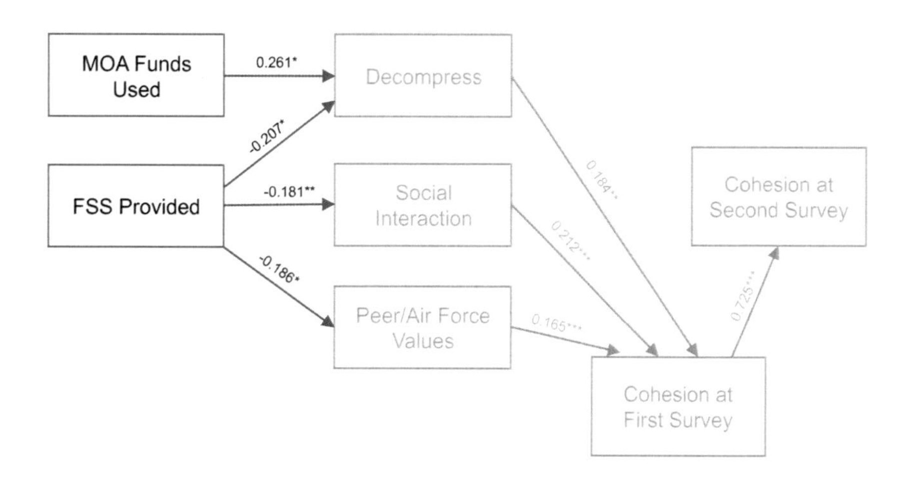

NOTE: *N* = 624. Only paths with significant associations are illustrated. The full model allows for all possible paths between event characteristics, building blocks, and cohesion outcomes, as well as paths between building blocks and cohesion from the second post-event survey. Standard errors (not shown) are clustered at the installation level. Models control for gender, age, rank, remote or isolated installation, and installation size. * indicates $p < 0.05$; ** indicates $p < 0.01$; and *** indicates $p < 0.001$.

How Are Readiness and Resilience Building Blocks Directly and Indirectly Associated with Overall Unit, Social, and Task Cohesion at Both the First and Second Post-Event Surveys?

Research question 2 asks how the readiness and resilience building blocks are directly and indirectly associated with overall unit, social, and task cohesion at both the first and second post-event surveys. Figure 4.3 illustrates the path analysis results pertinent to this research question, focusing on overall unit cohesion.

Figure 4.3. Path Analysis Results Related to Research Question 2, Overall Unit Cohesion

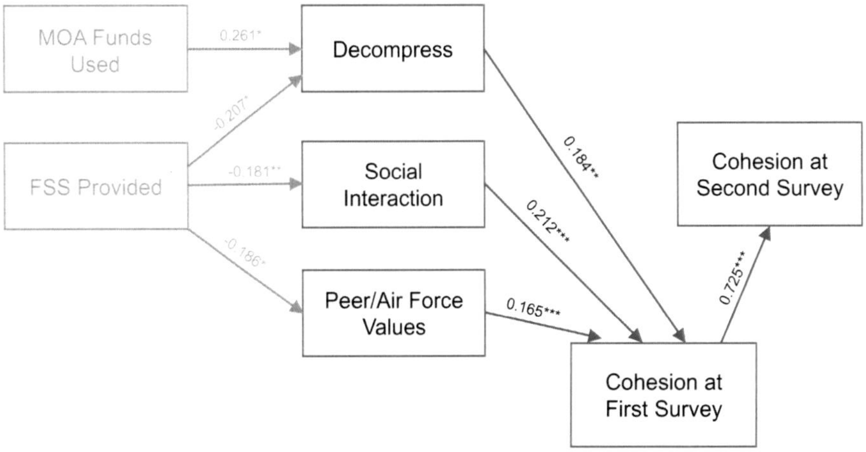

MOA Funds Used → Decompress: 0.261*
FSS Provided → Decompress: -0.207*
FSS Provided → Social Interaction: -0.181**
FSS Provided → Peer/Air Force Values: -0.186*
Decompress → Cohesion at First Survey: 0.184**
Social Interaction → Cohesion at First Survey: 0.212***
Peer/Air Force Values → Cohesion at First Survey: 0.165***
Cohesion at First Survey → Cohesion at Second Survey: 0.725***

NOTE: $N = 624$. Only paths with significant associations are illustrated. The full model allows for all possible paths between event characteristics, building blocks, and cohesion outcomes, as well as paths between building blocks and cohesion from the second post-event survey. Standard errors (not shown) are clustered at the installation level. Models control for gender, age, rank, remote or isolated installation, and installation size. * indicates $p < 0.05$; ** indicates $p < 0.01$; and *** indicates $p < 0.001$.

The direct associations between the building blocks and overall unit cohesion from the first post-event survey indicate that decompression, social interaction, and peer/Air Force values were positively associated with overall unit cohesion, measured two weeks or sooner after UNITE participation. The largest association was with social interaction, such that a one standard deviation increase in social interaction was associated with a 0.212 standard deviation increase in overall unit cohesion at that time ($p < 0.001$). The second-largest association was with decompression, such that a one standard deviation increase in the decompression building block was associated with a 0.184 standard deviation increase in overall unit cohesion at that time ($p < 0.01$). Finally, the lowest association was with peer/Air Force values, such that a one standard deviation increase in the values building block was associated with a 0.165 standard deviation increase in overall unit cohesion at that time ($p < 0.001$). We found no significant direct associations between the building blocks and overall unit cohesion rated by airmen at the second post-event survey.

To understand the total associations between the building blocks and overall unit cohesion from the second post-event survey, we must calculate the combined association of both the direct and indirect (i.e., building blocks via cohesion from the first post-event survey) pathways. Table 4.5 shows the total associations for the building blocks and overall unit cohesion at the second post-event survey. What we find is that the total associations for the decompression and peer/Air Force values building blocks with overall unit cohesion from the second post-event survey are very similar to the associations with overall unit cohesion from the first post-event survey. For example, a one standard deviation increase in the decompression building block is associated with a 0.185 standard deviation increase in overall unit cohesion at the second time point ($p <$

0.001). However, the total association between the social interaction building block and overall unit cohesion from the second time point is about one-third as large as the association with overall unit cohesion at the first post-event survey and is no longer significant. Thus, although the associations between the decompression and peer/Air Force values building blocks persisted to overall unit cohesion at the second survey, the association between social interaction and overall unit cohesion faded. Appendix D shows that the fading of this association was driven by social, rather than task, cohesion.

Table 4.5. Total Associations of Building Blocks with Overall Unit Cohesion at Second Post-Event Survey

Building Block	Cohesion at Second Post-Event Survey
Decompression	0.185***
	(0.053)
Social interaction	0.071
	(0.062)
Peer/Air Force values	0.148**
	(0.051)

SOURCE: RAND analysis of airmen survey data.
NOTE: $N = 624$. Standard errors in parentheses are clustered at the installation level. Models control for gender, age, rank, remote or isolated installation, and installation size. ** indicates $p < 0.01$; *** indicates $p < 0.001$.

How Are Event Characteristics Associated with Overall Unit, Social, and Task Cohesion Both Directly and Indirectly Through the Readiness and Resilience Building Blocks?

Our last research question asks how event characteristics are associated with overall unit, social, and task cohesion both directly and indirectly through the readiness and resilience building blocks. Figure 4.4 illustrates that all pathways are pertinent to this research question, focusing on overall unit cohesion. No arrows directly connect any event characteristic with overall unit cohesion from either the first or second post-event survey, indicating that the event characteristics are not directly related to overall unit cohesion at either time point.

Figure 4.4. Path Analysis Results Related to Research Question 3, Overall Unit Cohesion

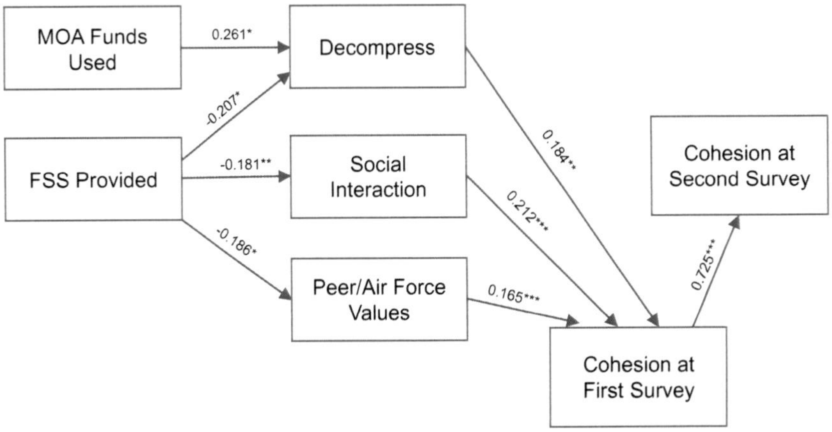

MOA Funds Used
0.261*
Decompress
-0.207*
FSS Provided
-0.181**
Social Interaction
-0.186*
Peer/Air Force Values
0.184**
0.212***
0.165***
Cohesion at Second Survey
0.725***
Cohesion at First Survey

NOTE: N = 624. Only paths with significant associations are illustrated. The full model allows for all possible paths between event characteristics, building blocks, and cohesion outcomes, as well as paths between building blocks and cohesion from the second post-event survey. Standard errors (not shown) are clustered at the installation level. Models control for gender, age, rank, remote or isolated installation, and installation size. * indicates $p < 0.05$; ** indicates $p < 0.01$; and *** indicates $p < 0.001$.

Estimating indirect and total associations requires calculating the combination of the pathways in Figure 4.4, and we present those results in Table 4.6. In regard to events for which MOA funds were used, there were no significant indirect or total associations with overall unit cohesion at either time point. In regard to events that were FSS-provided, there was a significant negative indirect association (through all building blocks) of −0.107 standard deviation ($p <$ 0.01) with overall unit cohesion at the first post-event survey, but that association fades to insignificance and is about one-third as large at the second post-event survey. Thus, the indirect effect fades over time. When taking into account all possible paths, there are no significant total associations between FSS-provided events and overall unit cohesion at either time point. Results were similar for task and social cohesion, with the negative indirect association between FSS-provided events and overall unit cohesion at the first post-event survey driven by task cohesion.

Table 4.6. Indirect and Total Associations Between Event Characteristics and Overall Unit Cohesion from First and Second Post-Event Surveys

	Cohesion at First Post-Event Survey		Cohesion at Second Post-Event Survey	
Event Characteristics	Indirect Association	Total Association	Indirect Association	Total Association
Used MOA funds	0.084	0.027	0.025	0.115
	(0.045)	(0.113)	(0.082)	(0.099)
FSS-provided	−0.107**	−0.051	−0.038	−0.048
	(0.035)	(0.076)	(0.056)	(0.086)

SOURCE: RAND analysis of airmen survey data.
NOTE: N = 624. Standard errors in parentheses are clustered at the installation level. Models control for gender, age, rank, remote or isolated installation, and installation size. ** indicates $p < 0.01$.

Quantitative Results Summary

Using path analysis, we were able to model the direct, indirect, and total associations between UNITE event characteristics, building blocks, and overall unit cohesion outcomes from the first and second airmen surveys. Results show that events provided by the installation FSS generally were associated with lower ratings of the building blocks but had no direct or total associations with overall unit cohesion at either time, and the only significant indirect association was a negative one with overall unit cohesion at the first post-event survey. Events that leveraged MOA funds had a positive association with ratings of the decompression building block but had no direct, indirect, or total association with any overall unit cohesion outcome.

The building blocks were more robustly associated with overall unit cohesion. Decompression, social interaction, and peer/Air Force values all were positively associated with overall unit cohesion at the first survey. Although they had no direct association with overall unit cohesion at the second survey, decompression and peer/Air Force values had a total significant and positive association with overall unit cohesion at the second survey. These results suggest that the association between decompression and peer/Air Force values and overall unit cohesion persisted from first to second post-event surveys, but the association between social interaction and overall unit cohesion faded over time.

Summary

In this chapter, we focused on the short-term and intermediate outcomes that UNITE aims to achieve, as summarized in the logic model presented in Chapter 2. We relied primarily on quantitative data from post-event airmen surveys. Using a path analysis, we found that two event characteristics were associated with how airmen viewed UNITE events. Specifically, events offered by the installation FSS were viewed as less of an opportunity to unwind and decompress, socially interact with peers, and promote peer, squadron, and Air Force values. Activities that leveraged MOA funds were also negatively (though indirectly) associated with overall unit cohesion two weeks after the UNITE event. Conversely, events that leveraged MOA funds were positively associated with decompression.

UNITE events that airmen rated as providing an opportunity to decompress, connect with their fellow airmen, and promote institutional values were positively associated with all three cohesion measures (i.e., overall unit, social, and task) immediately after participation (i.e., roughly two weeks). Although the association between decompression and peer/Air Force institutional values and overall unit cohesion remained eight weeks post-participation, the association with social interaction faded and became insignificant for overall unit cohesion, largely driven by a decline in social cohesion.

Several of the short-term and intermediate outcomes were also supported by qualitative survey data provided by UNITE participants. Airmen indicated that UNITE events provided an opportunity to build unit morale and cohesion, team-build, socialize and interact with other unit members, and relax, unwind, and decompress.

In the next chapter, we briefly summarize the overall results of our implementation and outcome evaluations of UNITE and offer recommendations for the Air Force to improve on its efficiency and effectiveness.

Chapter 5. Summary and Policy Implications

The Air Force, at the direction of Air Force Chief of Staff General David L. Goldfein, has made a concerted effort to revitalize the squadron (Goldfein, 2016). As Gen. Goldfein noted, squadrons are the "beating heart" of the Air Force. As part of this effort, the Air Force created an initiative, called UNITE, to provide unit commanders with discretionary funding to host events designed to improve unit cohesion. The Air Force also created a new position (C3s) to assist commanders in the planning and execution of these events. The ultimate goal of UNITE is to improve airmen and unit readiness and resilience via increased unit cohesion.

The Air Force asked RAND Project Air Force researchers to evaluate the UNITE Initiative. The research team developed an evidence-informed framework that built on an earlier study (Meadows et al., 2019), which outlined a set of building blocks for airmen readiness and resilience. Using this framework, the team developed a logic model for the UNITE Initiative (see Figure 5.1), which became the roadmap for the evaluation. Using both quantitative and qualitative data from C3s, unit commanders (or, in some cases, their designated POCs), and airmen who participated in UNITE events, we developed a set of analyses that examined the elements of the logic model: resources and inputs, activities, outputs, short-term outcomes, and intermediate outcomes. Although the analyses did not assess the ultimate impact of UNITE on airmen and unit readiness and resilience, we discuss this issue later in the chapter.

Figure 5.1. UNITE Logic Model

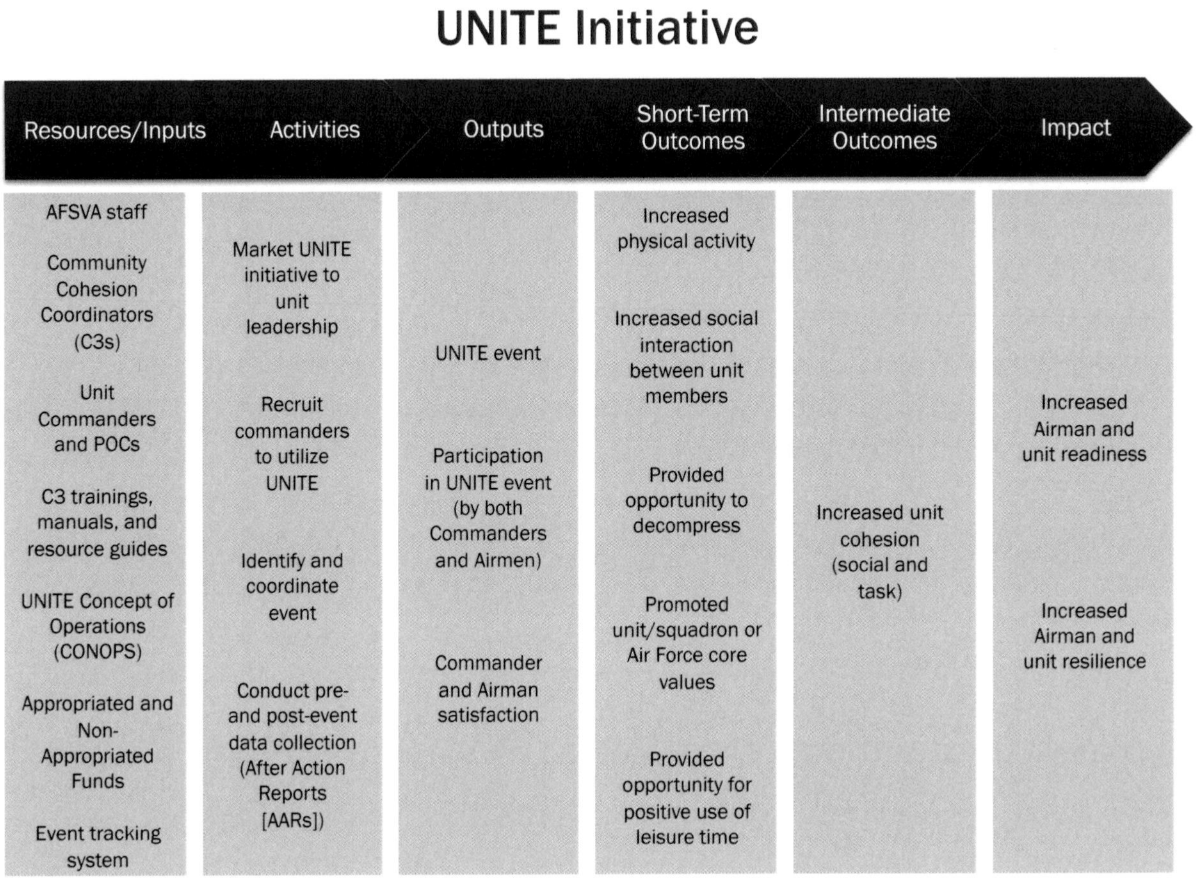

UNITE Initiative

Resources/Inputs	Activities	Outputs	Short-Term Outcomes	Intermediate Outcomes	Impact
AFSVA staff	Market UNITE initiative to unit leadership	UNITE event	Increased physical activity	Increased unit cohesion (social and task)	Increased Airman and unit readiness
Community Cohesion Coordinators (C3s)	Recruit commanders to utilize UNITE	Participation in UNITE event (by both Commanders and Airmen)	Increased social interaction between unit members		Increased Airman and unit resilience
Unit Commanders and POCs	Identify and coordinate event	Commander and Airman satisfaction	Provided opportunity to decompress		
C3 trainings, manuals, and resource guides	Conduct pre- and post-event data collection (After Action Reports [AARs])		Promoted unit/squadron or Air Force core values		
UNITE Concept of Operations (CONOPS)			Provided opportunity for positive use of leisure time		
Appropriated and Non-Appropriated Funds					
Event tracking system					

NOTE: AFSVA = Air Force Services Agency.

Before reviewing the results of the implementation and outcome evaluations, a few words of caution are warranted. Because UNITE was a new initiative when we began our study, it was constantly evolving as our evaluation took place. For example, we heard from C3s that units were frustrated about the timing of when they could (and could not) use funding that was allocated from two different resources, with two different expiration dates (e.g., one tied to the end of the calendar year, and one tied to the end of the fiscal year).[28] We also heard that the types of activities that were approved for funding by AFSVC felt like a moving target to unit commanders and POCs. At its inception, UNITE funding was only available to active-duty airmen (both active component and activated reserve component personnel) but was later expanded to include the Total Force. Implementation, and rules and regulations about implementation, also matured as the initiative was rolled out. Therefore, the results of our implementation analysis may have been influenced by the changing nature of UNITE.

[28] During the writing of this report, the research team learned that, starting in calendar year 2020, all funding sources would be tied to the calendar year.

Our outcome evaluation should be seen as a proof of concept, primarily because we could not control where and when UNITE was rolled out across the Air Force. Therefore, we were not able to conduct a randomized control trial (RCT). RCTs are the gold standard in program evaluation because the fact that they are randomized allows researchers to say that a program or intervention caused an observed result. We could not randomly assign airmen (or installations) to use UNITE, nor were we able to compare participants and nonparticipants, because the Air Force's goal was to get all units to use the funding and hold events. Instead, we developed a post-participation-only analysis, collecting data from UNITE participants after they attended a UNITE event. Thus, we are unable to claim that our analysis is a reflection of a causal relationship between UNITE and measured outcomes—that is, that UNITE *caused* a change in the building blocks or cohesion. Instead, we are limited to the conclusion that certain UNITE event characteristics were associated with post-event measures of the building blocks and cohesion.

Implementation

Resources and Inputs

C3s were viewed as a valuable resource by unit commanders and POCs, though C3s sometimes felt overburdened by a combination of UNITE and other duties assigned to them by installation leadership. They also showed interest in continued training and education, as well as more networking with fellow C3s.

One area in which we heard consistent criticism was the inconsistent and ever-changing guidance provided by AFSVC: particularly, what was and was not covered by UNITE funding. Various types of informants also noted that increasing the amount of funding that could be used on food could increase both participation in and satisfaction with UNITE events. Assignment of unit POCs and turnover of leadership were two other potential areas that may have a negative impact on the implementation of UNITE.

Not all C3s were completely satisfied with current data collection mechanisms, such as SharePoint. Understanding what data is needed by various stakeholders (e.g., financial data, event data, outcome data), together with feedback from current system users, could inform changes to these systems that could make them more effective and more efficient.

Activities

Messaging about UNITE in terms of simply making sure that airmen are aware of specific events and that they know the purpose of the overall UNITE Initiative were two areas in which C3s, unit commanders, and airmen all agreed that improvements could be made. One cause of concern is the perception by some airmen that UNITE events are "mandatory fun." Another is the lack of consistent marketing materials available to C3s.

C3s used a variety of methods to work with units in developing and planning activities. One of the most ubiquitous complaints voiced by C3s, unit commanders, and POCs was the requirement to plan in advance: specifically, the time it takes to get a proposed UNITE event approved by AFSVC. Many units expressed that they would like more freedom over how UNITE funds are used.

Data collection about UNITE events is another potential area of concern. As noted earlier, current online data collection and collaboration tools were not seen as efficient or effective by some C3s. We also observed problems with the collection DoD ID numbers of participants, which had implications for our outcome evaluation.

Outputs

Overall, airmen and commanders indicated a high degree of satisfaction with how UNITE was implemented. Many of the comments we received suggested that airmen, in particular, perceived UNITE events to be an opportunity to relax and unwind while getting to know their fellow unit members. However, we also heard complaints about low attendance at some events, which could mitigate some of the perceived positive effects of the initiative.

Outcomes

Short-Term Outcomes

The short-term outcomes in the UNITE logic model focus on a select group of readiness and resilience building blocks. Path models revealed that UNITE events offered by the installation FSS were viewed by airmen as less of an opportunity to unwind and decompress, socially interact with peers, and promote peer, squadron, and Air Force values. Conversely, MOA-funded events were positively associated with decompression. None of the event characteristics were significantly associated with physical activity.

Intermediate Outcomes

Participation in UNITE events that airmen rated as an opportunity to decompress, interact with other airmen, and reinforce peer, squadron, and Air Force values were associated with higher levels of overall unit cohesion two weeks after participation. However, the associations between social interaction and overall unit cohesion dissipated over time (i.e., eight weeks after participation).

Policy Implications

Implementation of UNITE

Although there are numerous positive aspects of UNITE (most notably, the overwhelmingly positive response by unit commanders and airmen), there are several areas in which changes could be made to the implementation of the initiative that could result in improved efficiency and effectiveness. These recommendations are especially relevant to AFSVC because it supports installation-level FSSs. Next, we describe a few of these changes.

Increase awareness of UNITE among unit commanders. Commanders and airmen noted that they were not aware of UNITE until they heard of other units having events. C3s noted that one of their biggest obstacles to successfully implementing UNITE was a lack of awareness on the part of units. First and foremost, AFSVC should consider providing each C3 with a marketing budget that, like other UNITE funding, could be based on the number of airmen at a given installation or the number of eligible units. In addition, creating marketing templates or marketing packages for C3s to adapt to their installations would make marketing more consistent across the Air Force.

Improve UNITE messaging through use of standardized materials. We heard from C3s and airmen that the purpose of UNITE is not always obvious. Unit commanders should be clear that UNITE events are designed to enhance unit cohesion, often through team-building. Boosting morale and providing an opportunity to relax or rejuvenate are added benefits, but these shorter-term outcomes should be seen as a pathway to achieving increased cohesion, which is the primary purpose of UNITE. In addition, when UNITE activities are combined with other programs (e.g., suicide prevention stand-downs or resilience days), the purpose of UNITE can be obscured.

Develop other materials and processes that may increase uptake of UNITE. Developing other materials in addition to the marketing materials could help C3s make UNITE easier for unit commanders to use. First, develop a set of tools to structure the C3 and unit commander or POC interactions. These tools could include introductory email templates, flyers, and training binders for C3s to use with POCs. Some C3s have already developed such tools, and they could serve as drafts for others to adapt. Second, develop a set of UNITE RTE options that could be preapproved and used by units on short notice. These would be events that have been approved in the past and are already known to meet UNITE guidelines. Having such a list of on-demand activities could make it easier for units that are limited in the amount of time they can be away from their workspace or that have unpredictable schedules to also take advantage of UNITE.

Provide C3s with the tools and resources they need to be effective in their jobs. According to our interviews with C3s, we learned that some of them are using personal funds to cover expenses incurred when fulfilling their duties, including when paying for marketing materials, cell phones, and transportation. As noted earlier, standardized marketing materials would, to some degree, centralize this task and take the burden off C3s, especially if they are already paying for such materials on their own. Although it may not be feasible to provide every C3 with a vehicle, they could be reimbursed for mileage when travel is required (e.g., to pick up supplies for a UNITE event or to scout out new venues). Providing them with a dedicated work cell phone could be feasible, however, and would allow C3s to use such phones for UNITE purposes only (e.g., as an emergency number in case something goes awry during an event). Some C3s described administrative hurdles to executing events, such as the paperwork required before completing a transaction with a P-card. To the extent that such processes can be streamlined, C3s will have more time for substantive aspects of their work and may be able to more quickly meet the needs of units who would like to plan a quick-turnaround event. We also heard that some C3s are "borrowed" by installation leadership to cover other activities or programs. If possible, C3s should have protected time devoted only to UNITE duties. Finally, AFSVC should consider whether having additional staff in place (including a second C3) at larger installations could alleviate some of the time constraints that C3s at such installations are currently facing (e.g., not being able to check on UNITE events in person). Some larger installations have explored options for providing C3s with support, though this was only beginning to be implemented at the end of our evaluation period.

Improve data collection tools, making them more user-friendly and more consistent. Reviews of existing data collection and collaboration tools were mixed, with some C3s having real difficulty and others being ambivalent about their experience. As users of the data collected by some of these tools, we believe that improvement is needed. The event tracking system, which we have referred to in this report as the C3 UNITE event database on SharePoint, should be more user-friendly, with clear guidance on the type of data that should be inputted into each field. AAR forms should also be user-friendly, with consistent guidance on exactly what information should be captured. Consistency in the data captured by the event database and AARs is the only way to ensure that these data can be used for a high-quality evaluation of UNITE. C3s should also easily be able to track events and access information about past and future activities, if needed. Finally, the event request process should be simplified and streamlined.

Hold annual, in-person refresher training for C3s. Although C3s were generally pleased with the initial in-person training they received in San Antonio, Texas, several mentioned that annual refresher training would be helpful. This training would serve two purposes. First, it would allow C3s to learn about any new procedures or processes put into place by AFSVC and allow for real-time feedback if C3s have concerns or questions. Second, it would allow for face-to-face interaction with other C3s and provide an opportunity to share effective practices for planning events with one another. Annual refresher training would not take the place of existing collaboration tools (e.g., Facebook, Blackboard) but would supplement it. Such additional training may be especially beneficial for C3s who are currently reluctant to participate in existing forms of social networking because of a belief that they are impersonal and not useful.

Provide clearer and more consistent guidance on UNITE policies, processes, and procedures. C3s perceived the guidance they received from AFSVC to be unclear and inconsistent, which may reflect that UNITE was evolving during the period in which we conducted our evaluation. Nonetheless, AFSVC needs to provide clear and consistent guidance about the policies, processes, and procedures surrounding UNITE. This includes such topics as the nature of the different sources of funding, how they can and cannot be used, and specific elements that must be provided in a request form for an event to be approved. We are only aware of one guidance document: the UNITE CONOPS. Thus, it is worth considering a more formal guidance document, such as an AFI, that would codify policies, processes, and procedures that appear to C3s to be piecemeal. Finally (as noted earlier), any changes to UNITE guidance should be clearly communicated to relevant stakeholders, especially C3s, through multiple mechanisms, including existing networking tools (e.g., Facebook, Blackboard) and in-person refresher training.

Consider where UNITE should be structurally located within the Air Force to best position it to achieve its broader goal of improving airman and unit readiness and resilience. Currently, there is no requirement for UNITE events to use FSS programs or services. Therefore, it may be useful to consider whether UNITE should reside in AFSVC. AFSVC's mission statement does not include language about cohesion, readiness, or resilience.[29] The Air Force has several options for locating UNITE, should it decide to move it. It could be aligned with the personnel management function of the Air Staff (AF/A1). AF/A1 has both a Directorate of Services (AF/A1S) and an Integrated Resilience Directorate (AF/A1Z). Placing UNITE in AF/A1Z would reinforce the fact that, ultimately, UNITE is about improving airman and unit readiness and resilience. One C3 pointedly noted that UNITE's placement within the FSS flight limited their ability to implement the initiative because it was harder for them to reach wing commanders. Being at the wing level, where the Resiliency Program resides, was seen as a source of authority.

Effectiveness of UNITE

We also identified potential opportunities to increase the effectiveness of UNITE. These recommendations are directly relevant to installation FSSs, C3s, and commanders who are planning UNITE events.

Repeat UNITE events throughout the year. Results from the path models show that events where airmen could relax and unwind, socially interact with peers, and reinforce peer, squadron, and Air Force values were positively associated with overall unit cohesion immediately after participation (i.e., roughly two weeks after). Just eight weeks after participation, the association with social interaction had faded to insignificance, and the other associations were reduced in magnitude. Further, most of the associations between the building blocks and overall unit

---

[29] See U.S. Air Force, 2002.

cohesion at the second survey were driven by a very strong association between overall unit cohesion at the first and second surveys. The diminishing effects of UNITE are probably not surprising and suggest that certain types of events—those that allow for decompression, fellowship, and celebration of what it means to be an airman—should be repeated. Our results do not suggest how often such events should occur, though this is an empirical question that could be answered with a different UNITE implementation and research evaluation plan (see the following section on next steps). Further, it may not be financially viable for units to hold more than one UNITE event per funding cycle. Emphasizing volunteer events (e.g., building a Habitat for Humanity house, manning a food bank or kitchen) could provide one way to stretch UNITE dollars further and allow for multiple events throughout the year.

Emphasize the importance of events with a particular focus. Earlier, we noted the importance of events that targeted specific building blocks: coping skills and strategies (i.e., opportunity to decompress, consistent with the leisure as coping literature), social network (i.e., opportunity to interact with fellow airmen), and peer/squadron and Air Force values (i.e., reinforcing and promoting shared beliefs). Knowing that these building blocks might make a difference can help serve as a blueprint of sorts when designing events. This may mean that events that are double-billed as UNITE and a resilience day or a suicide stand-down day may not be as effective as events that target only decompression, social interaction, or reinforcing shared values. This recommendation is also consistent with an earlier recommendation to make sure the goal of UNITE events is more explicit: They are not just fun barbeques, the event is about building cohesion through shared downtime, socializing, team-building, and reinforcing the Air Force way of life together.

Take advantage of opportunities to hold UNITE events outside the gates of the installation. One particularly intriguing finding from our path models was the negative association between activities provided by the installation FSS and the building blocks. We know that the overwhelming majority of FSS-supported events were held on base (92 percent). Moreover, qualitative data from participants support the notion that off-base events were preferred. Together, these data suggest that airmen were more receptive to events that took them away from day-to-day life on the installation. Curiously, however, we found no significant associations between off-base events and any of the building blocks or overall unit cohesion at either time point. Thus, although the FSS variable in the model may be absorbing some of the association between event location and the building blocks and cohesion (given the high correlation between the two), the location of the event alone is unlikely to fully explain the association we observed. We attempted to do a more in-depth analysis of the event description data from AARs by examining spending levels and descriptions of activities that were part of UNITE events. Unfortunately, because of the inconsistent nature and poor quality of these data, our exploration did not yield useful insights into why FSS-sponsored events were negatively associated with the building blocks.

Next Steps

The implementation and outcome evaluations of UNITE presented in this report are only a first step toward assessing whether UNITE is achieving its goals of improving unit cohesion and airmen and unit readiness and resilience. Should the Air Force decide to continue the initiative, there are several steps it could take to further its evaluation efforts.

Design and implement a more rigorous evaluation plan. We noted in Chapter 4 that the optimal research design for an evaluation of UNITE would compare cohesion among airmen who did and did not participate in a UNITE event. Alternatively, a before-and-after participation approach would be a close second in terms of methodological rigor. Unfortunately, neither approach was an option, given the constraints of this study. Moving forward, however, both approaches are possible with enough planning time.

Improve the quality of data collected for future evaluation efforts. Once a more rigorous evaluation plan is developed, it will require high-quality data. Much of the date we relied on for our analysis was inserted by hand into less-than-user-friendly databases and software. Any time data is transferred from a subject to a holding place—such as a database—by a third party, the odds of errors in data translation increase. C3s told us that units had to collect UNITE participants' DoD IDs and input them into existing data systems by hand. DoD IDs are ten-digit numbers that can easily be transposed. Providing C3s with handheld Common Access Card readers that units can use to track participation would eliminate many of the data quality issues that UNITE faces. Similarly, participation data need to be directly linked to UNITE events using unique event numbers that cannot be reused. Finally, participation data is not the only type of data that should be high quality. It is also very important to collect standardized data about UNITE events themselves. Such data might include the location of the event, partnerships with specific vendors, the time of day, the reason for the event, and more-detailed information about funding. The existing event database might need to be scrubbed and revised, and explicit instructions for what data to submit (and how to submit the data) might need to be provided to C3s, unit commanders, and POCs. Having such standardized, detailed event information would help to untangle the reason we observed a negative association between FSS-sponsored activities, building blocks, and cohesion. An additional benefit associated with improved data collection capabilities and the higher-quality data they will produce is that evaluation-related questions can become more nuanced. For example, what is the ideal size of a UNITE event? A squadron event may be too large, but is a workstation too small? What impact does inviting family members have? Are there certain types of activities (e.g., competitions) that are more strongly associated with some building blocks and ratings of unit cohesion?

Assess the impact of UNITE on airmen and unit readiness and resilience. At the inception of this study, we explored ways to measure readiness and resilience at the individual and unit levels, including by drawing on administrative data from a variety of sources. This proved more difficult and more resource-intensive than expected and could not be included in our first-stage evaluation of UNITE. However, in the future, the Air Force may wish to pursue risk and resilience profiles of airmen that could be aggregated at the unit level. For example, these profiles could include negative health behaviors, disciplinary actions, family issues, and financial problems. They could be collected from existing administrative and survey data (assuming they could be linked to individual airmen). The data would never be reported at the individual level, however, allowing airmen's personal information to remain confidential. Such profiles could provide an overall picture of how well a given unit is functioning and how prepared they may be for an upcoming deployment, training, or other stressful event.

Conclusion

With this study, we conducted an initial evaluation of the UNITE initiative. Our findings suggest that the initiative has been well-received by airmen, though there may be opportunities to improve the efficiency and effectiveness of implementation. In addition, we found that UNITE events appear to target building blocks of resilience and readiness that have been identified in the literature; in turn, these building blocks were associated with perceptions of overall unit cohesion following the event, though these associations faded over time. These findings emphasize the importance of continuing to evaluate UNITE to better understand its impact on individuals and units, especially given that the literature on interventions to promote cohesion is limited.

Appendix A. Cohesion Literature Review Methods and Findings

We conducted a review of the cohesion literature to understand and identify the predictors of group cohesion. For the purpose of this review, we focused on cohesion within a single group, such as sports teams or military units. A separate stream of social cohesion literature, predominantly in the disciplines of sociology and political science, focuses on the prevalence of "common values and a civic culture; social order and social control; social solidarity and reductions in wealth disparities; social networks and social capital; and territorial belonging and identity" (Kearns and Forrest, 2000, p. 996). In that context, social cohesion focuses on the "connectedness and solidarity" across numerous groups within a society (Kawachi and Berkman, 2000). For this study, however, we directed our attention to the study of within-group cohesion, because this is more applicable to understanding cohesion within Air Force units. In this appendix, we describe our literature search strategy and findings.

Overall Literature Search Strategy

We conducted a targeted search of the scientific literature on group and unit cohesion. Our goal was twofold. First, we aimed to identify predictors of group cohesion. Second, we aimed to identify interventions designed to promote or develop group cohesion.

We began by searching the following databases: Web of Science, PsycINFO, and PubMed. To identify correlates of group cohesion, we restricted the search to include articles that have the terms "social cohesion," "group cohesion," or "unit cohesion" in the manuscript title.

We then conducted additional targeted searches for research that examined small-group cohesion interventions, building on the studies identified in our search for cohesion predictors. Specifically, we sought research that conducted primary analysis on a small-group intervention focused on improving cohesion. We concentrated the focus of our intervention studies review on understanding (1) the quality of research evidence for cohesion interventions and (2) the mechanisms through which these interventions are expected or suggested to increase cohesion. We describe our definition and framing of these two components later in this appendix. We searched for cohesion intervention evaluations in the following databases: Web of Science, PsycINFO, and PubMed. We searched for manuscripts that contained the terms "social cohesion," "group cohesion," or "unit cohesion" in the title or abstract and the terms "intervention or treatment or program or strategy" in the text.

While reading the cohesion intervention literature, we learned that the term "team building" was used instead of "group cohesion" when discussing such interventions. We conducted a supplemental literature search to identify research that was relevant to group cohesion interventions within the team-building literature. We searched Web of Science, PsycINFO, and

PubMed for articles that had the term "team building" in the title or abstract and the terms "intervention or treatment or program or strategy" in the text.

Finally, to ensure that our search strategy was comprehensive and identified the desired body of research, we had a researcher run an independent search on terms related to cohesion and team-building. Specifically, this researcher scanned Business Source Complete, PsycINFO, PsycArticles, Google Scholar, and RAND's library catalog for research. The search was restricted to academic journals, magazines, and trade journals for articles that had at least one term from each of the following groups in either the title or the abstract:

- unit cohesion, group cohesion, military cohesion, peer bonding, teamwork, work group cohesion, team building, task cohesion, group environment questionnaire, resilience, trust, or mental health
- training, program, intervention, evaluation, evaluate, experiment, trial, control group, or comparison.

Review of Cohesion Predictor Literature

The search for cohesion correlates identified 1,368 citations. After removing duplicate citations, manuscripts published in a language other than English, and nonpeer-reviewed literature (e.g., dissertations, books), 761 eligible citations remained. We then conducted a two-phase screening process to select studies that (1) examined human behaviors (there were a large number of studies about animal behaviors), (2) assessed "social cohesion," "group cohesion," or "unit cohesion" as an outcome, and (3) investigated cohesion within small groups (e.g., sports teams, work units, military units). Small-group setting was included as a selection criterion because it is the most relevant setting to the Air Force squadron or unit, the focus of this study. First, two members of the project team conducted a title and abstract review; articles identified as relevant during this process underwent a full-text review. A third team member reviewed any discrepancies. After the title and abstract review, 160 manuscripts were screened as potentially meeting our inclusion criteria. Eighteen manuscripts focused on interventions to promote cohesion; these were flagged for consideration in the cohesion intervention literature review (described in more detail later in this appendix). After completing the full-text review on the remaining manuscripts, 59 manuscripts were found to be available and relevant. Each included study was reviewed by two members of the research team, and relevant predictors of cohesion were abstracted. Any discrepancies were reviewed by a third member of the team.

We included both qualitative and quantitative studies in our review. If the study used qualitative methods, we included a definition of cohesion in our abstraction. If the study used quantitative methods, we extracted the measure(s) of cohesion and identified the analytic approach used to produce the evidence. From each manuscript, we extracted (1) the study population (i.e., civilian or military), (2) the definition or measure of cohesion, (3) the study design, (4) the predictors of cohesion, and (5) the results.

The majority of these predictors were identified through cross-sectional research designs. Only eight manuscripts presented research using experimental or quasi-experimental designs. Therefore, findings from the literature review do not necessarily provide evidence to suggest a causal relationship between the identified predictors and cohesion. However, the predictors are guided by theory and expected to increase cohesion.

As a final step, the two project leaders reviewed the abstracted predictors and categorized them into larger domains, which are presented next.

Predictors of Group Cohesion

In this section, we present the predictors of within-group cohesion according to our literature review. Table A.1 indicates each higher-level category of predictor (e.g., group communication, leadership) and provides a description of the findings related to each predictor. There are certain caveats to this literature. In most cases, cohesion was measured at the individual level (i.e., individuals were asked to rate their perceptions of group cohesion) rather than at the group level. In addition, many of these studies were cross-sectional, measured at one point in time, which precludes inferences regarding the causal association between each predictor and cohesion. However, this review demonstrates the variety of factors that are associated with cohesion.

Table A.1. Group Cohesion Predictors

Cohesion Predictor	Description of Predictor's Association with Group Cohesion
Demographic: social characteristics	• Includes such characteristics as the gender composition, age of participants, race/ethnicity, and marital status. • Evidence of **age's** association with group cohesion has been mixed, with some studies showing a positive association (i.e., groups with older members report higher levels of cohesion; Griffith, 1989), but other studies have found no association between a group's age diversity and its level of cohesion (e.g., Harrison Price, and Bell, 1998; Lee and Farh, 2004; and Webber and Donahue, 2001). At least one study reported that greater age gaps among team members was associated with lower social cohesion (Aubke et al., 2014). • There have also been mixed results for **gender**, with studies finding that groups with a higher proportion of women have greater cohesion (Dermatis et al., 2001; Sanchez and Yurrebaso, 2009), at least one study finding that military units with a higher percentage of women experienced lower cohesion (Rosen et al., 1999), and others finding no effect of gender (Brisimis, Bebetsos, and Krommidas, 2018). • A study of Army soldiers found that units with a higher proportion of nonwhite soldiers had higher cohesion (Griffith, 1989), though other studies found no effect of **racial/ethnic diversity** (e.g., Harrison, Price, and Bell, 1998). • A study of military veterans also found that units whose individuals had a history of trauma or adverse childhood events had lower cohesion (Grady et al., 2018).
Demographic: work/military	• Includes work-related history, such as length of membership in a group, length of employment, and diversity of work-related experience. • Longer **time in a unit** is associated with greater cohesion among both military members and civilians (Bartone et al., 2002; Sanchez and Yurrebaso, 2009; Vaitkus and Griffith, 1990). • **Length of employment** or **length of time in a program** has mixed support: Some studies have found higher cohesion when employees have been employed longer (Aubke et al., 2014; Ko, 2011), though at least one study found the opposite (Sanchez and Yurrebaso, 2009).
Group communication	• A small number of studies examined the effect of group communication. • A study in college students found that individuals attending **in-person** classes perceived greater cohesion than those in online courses did (Galyon et al., 2016). • Another study in college students found that project groups had greater cohesion when members communicated more **effectively** (i.e., communicated in a way that achieved the intended goals) and more **appropriately** (i.e., consistent with social norms; Troth, Jordan, and Lawrence, 2012).
Healthy social interaction	• Includes such practices as promoting empathy and allowing for emotional expression. • A study of individuals participating in a therapy group found that when the leader promoted **expression of emotion**, encouraged **openness** among group members, and validated participant **self-expression**, greater group cohesion was achieved (Plante, 2006). • Groups with a greater **expectation of social interactions** among group members have been shown to have higher social cohesion (Benson, Eys, and Irving, 2016), and **risk-taking** in groups also appears to promote cohesion (Stokes, 1983).

Cohesion Predictor	Description of Predictor's Association with Group Cohesion
Individual subjective cognitions	• Includes perceptions of the work (e.g., job satisfaction), perceptions of the group (e.g., connection with group members, value of the group), and degree of identification with the group or its goals. • Evidence regarding the role of **job satisfaction** has been mixed, with some studies finding that greater satisfaction is associated with higher cohesion (Harrison, Price, and Bell, 1998) and others finding it is associated with lower cohesion (Ko, 2011). • The extent to which individuals **identify with the group** (López et al., 2015) and the greater organization (García-Guiu, Molero, and Moriano, 2015) appears to increase cohesion. However, **support for the group's mission** (Rosen et al., 1999) or perceived importance of group policies (Kim et al., 2016) do not appear to be associated with cohesion. • Perception of the **value or benefit of a group** is associated with greater cohesion (Dermatis et al., 2001; Stokes, 1983). • One study examined the **discrepancy between group members' ideal group culture and perceived actual group culture** and found that a larger difference between ideal and actual culture was associated with lower cohesion (Sanchez and Yurrebaso, 2009).
Leadership	• Includes both the characteristics and behaviors of leaders in promoting group cohesion. • When leaders are perceived as **authentic** (García-Guiu, Molero, and Moriano, 2015; López et al., 2015) and **providing social support, training and instruction, and positive feedback** (Jowett and Chaundy, 2004; Lee and Farh, 2004; Shields et al., 1997), members of their group endorse higher group cohesion. • Leaders who are perceived as **fair, honest, and transparent in decisions** also promote group cohesion (Charbonneau and Wood, 2018; García-Guiu, Molero, and Moriano, 2015; Ismail, Baki, and Omar, 2018). By contrast, **autocratic behavior** has been shown to be associated with poorer cohesion (Shields et al., 1997). • When there is a greater discrepancy between **leader perceptions of their leadership style** and the perceptions of group members, group cohesion may also be lower (Shields et al., 1997).
Mental health	• A small number of studies reported on the association between mental health symptoms and cohesion. • In a military sample, **posttraumatic stress disorder** and **depressive symptoms** were found to be negatively associated with unit cohesion (Welsh, Olson, and Perkins, 2019). • A study of civilians receiving care through a therapeutic community also found that **depressive symptomatology** was associated with lower cohesion (Dermatis et al., 2001).
Personality	• Regarding the Big Five personality traits, **agreeableness** and **conscientiousness** have been positively associated with cohesion, though the evidence has been mixed for **extraversion**, and **neuroticism** may lead to poorer cohesion (Aeron and Pathak, 2016; van Vianen and De Dreu, 2001). • Research related to **attachment style** has also been mixed. For example, Tiryaki and Çepikkurt, 2007, finds that secure and preoccupied attachment styles were positively associated with social cohesion, but preoccupied attachment styles were negatively associated with task cohesion. Others have found no association between attachment style and cohesion (Grady et al., 2018). • Other aspects of personality that are associated with higher cohesion include higher levels of **emotional intelligence** (Moore and Mamiseishvili, 2012; Troth, Jordan, and Lawrence, 2012), greater **emotional stability** (van Vianen and De Dreu, 2001), **hardiness** (Bartone et al., 2002), and a **need for inclusion** (Bugen, 1977).

Cohesion Predictor	Description of Predictor's Association with Group Cohesion
Group culture	• Certain aspects of group culture appear to promote group cohesion, including having a **noncompetitive working environment** (Rohe et al., 2006) and **task-oriented team climate** (Boyd et al., 2014). • By contrast, "development-oriented" organizations that are focused on **risk-taking**, **entrepreneurship**, **growth and resource acquisition**, and **rewarding individual initiative** may have lower group cohesion (Ismail, Baki, and Omar, 2018). • Similarly, research on college athletes found that **ego-involving group climate** (i.e., an emphasis on winning, recognizing the best players, and competition among teammates) was associated with poorer cohesion (Boyd et al., 2014).
Group dynamics	• The interpersonal dynamics among members of a group can influence group cohesion. For example, when group members need to **work with each other to achieve a goal** (e.g., because each member has a distinct set of responsibilities), cohesion has been demonstrated to be higher (Chen, Tang, and Wang, 2009; Turk, 1963). • **Group efficacy** (i.e., the extent to which a group believes in its combined abilities to work toward a goal) has been associated with cohesion in some studies (e.g., Heuzé, Bosselut, and Thomas, 2007) but not others (e.g., Lee and Farh, 2004). • Greater **cooperation**, **pride in the group**, and **organizational citizenship** (Kaymak, 2011; Lin and Peng, 2010; Pack and Rickard, 1975) have been associated with greater cohesion, and groups that work toward a collective goal also experience higher cohesion (Chen, Tang, and Wang, 2009).
Shared culture	• A small number of studies found that groups with shared culture—such as **shared attitudes, rituals, heritage, national pride, or religious identity**—have higher cohesion (Boer and Abubakar, 2014; Christensen et al., 2006; Nilsson, 2018).
Shared experience	• There is evidence that groups that have shared experiences have greater cohesion. The literature for this correlate fell into three categories. • **Prior group success**: Groups that have previously successfully achieved a goal or performed well together have been shown to have higher cohesion (Hoogstraten and Vorst, 1978; Kaymak, 2011; Lee and Farh, 2004). • **Neutral shared experiences**: Research has also shown that several types of shared tasks are associated with cohesion, including participating in interactive sports (Lafferty, Wakefield, and Brown, 2017), physical activity (Banning and Nelson, 1987; Jenkins and Alderman, 2011); and collaborative tasks, such as group story-building (Cordobés, 1997). • **Difficult shared experiences**: Studies with military populations have shown that sharing difficult experiences, such as basic training and deployment, may be associated with greater cohesion (Griffith, 1989; Vaitkus and Griffith, 1990), though another study found that combat experience was associated with lower levels of unit cohesion (Welsh, Olson, and Perkins, 2019).

Review of the Cohesion Intervention Literature

We identified 18 manuscripts on cohesion interventions through the cohesion predictors literature search, as described earlier. Eight of these manuscripts met the criteria for inclusion in our discussion of cohesion interventions following a full-text review. The remaining interventions come from additional searches of the cohesion intervention literature.

The first targeted search for cohesion interventions identified 1,191 citations. After removing duplicates, results published in a language other than English, and nonpeer-reviewed literature (e.g., dissertations), we identified 778 citations. We then conducted a title and abstract review to identify the studies that (1) examined human behaviors, (2) investigated cohesion within small groups (e.g., sports teams, work units, platoons), and (3) tested the effectiveness of a program and/or intervention that was designed to increase cohesion. After the title and abstract review, we identified 27 articles for full-text review, including five that were identified during the original cohesion predictors literature search. We conducted a full-text review of 22 manuscripts. After the full-text review, eight manuscripts met our inclusion criteria for the discussion of cohesion interventions evidence.

Our supplemental search of the team-building intervention literature returned 1,845 citations. After removing duplicate citations, manuscripts published in a language other than English, and nonpeer-reviewed literature (e.g., dissertations), 829 eligible manuscripts remained. We excluded 762 articles during the title and abstract review and removed two manuscripts for being duplicates of studies identified in our cohesion literature searches, leaving 67 manuscripts for full-text review. After the full-text review, 12 manuscripts unique to the team-building intervention search were included in our synthesis of cohesion intervention evidence.

Our final, independent search of the literature for evaluations of cohesion or team-building interventions resulted in 1,081 citations. After removing duplicate citations, manuscripts published in a language other than English, and nonpeer-reviewed literature, 999 eligible manuscripts remained. We excluded 934 of these manuscripts during the title and abstract review, leaving 65 manuscripts for full-text review. After the full-text review, one manuscript that was unique to this search was included in our synthesis of cohesion intervention evidence.

A total of 29 manuscripts, from across all of our literatures searches, were included in the review of group cohesion intervention.

Assessing Evidence Quality

The ultimate goal of evaluating interventions is to estimate the influence or effect of the intervention on the outcomes of interest (such as cohesion), accounting for all other factors that might affect the uptake of the intervention and the outcome. For example, when looking at how MWR programs affect unit cohesion, simply comparing the cohesion of squadrons that participated in MWR programs to those that did not would be unlikely to yield the true, unbiased estimate of the effect of the programs on the outcome of interest. This is because squadrons that voluntarily take up the MWR programs might be very different than those that do not. For example, unit leaders who sign up for the programs, as compared with those who do not, might be more conscious of their squadron's cohesion and might be engaged in other activities that increase cohesion. A simple comparison would then attribute the influence of these unobserved factors to the impact of participating in MWR programs, thus overstating the effect of the programs.

In light of this problem, researchers use a variety of techniques to account for these unobserved factors in an attempt to recover an unbiased estimate of the effect of an intervention on outcomes. Some techniques are more rigorous and more likely to uncover the unbiased estimate of the effect than others. The quality of evidence to support an intervention's contribution to cohesion is defined by the design and implementation of the analytic technique used to measure the relationship between the intervention and the cohesion outcome(s).

The most rigorous and highest-quality evidence is produced by *experimental designs*, in which participants are randomly assigned to the intervention or control condition. To be well-implemented, randomized studies should demonstrate low attrition (i.e., few individuals withdrawing from the evaluation) and should test for baseline equivalence in the two randomized groups on key attributes or preintervention outcomes of interest (e.g., a measure of cohesion before participating in UNITE). Results from experimental designs enable causal claims to be made; that is, the intervention is responsible for the identified changes in cohesion rather than some other unobserved or unmeasured influence. RCTs result in an unbiased estimate of the effect of the treatment or intervention.

Following experimental studies, *quasi-experimental designs* (QEDs) provide the next most rigorous evidence. These studies leverage an equivalent, or matched, comparison group against which the intervention participants can be compared. Well-designed and well-implemented QEDs demonstrate baseline equivalence between the intervention and comparison group. Where equivalence is not demonstrated, controls for these differences in the statistical analyses should be used to prevent sample bias from influencing the analytic results. QED results are suggestive of causal evidence, hinting at the *likely* causality between an intervention and identified changes in cohesion.

Correlational designs follow QEDs in rigor because there is no attempt in these approaches to leverage an equivalent comparison group. This class of statistical approaches can be further disaggregated into *correlational analyses with controls* or *correlational analyses without controls*. Analysis with controls accounts for differences in the individuals who receive the intervention compared with those individuals who did not receive the intervention. Similarly, this method will occasionally control for a measure of the focal outcome prior to the delivery of the intervention. This approach is more rigorous than the correlational analysis without controls because the latter approach does not account for sample selection or sample bias. Regardless of whether the correlational analyses use controls in the statistical analyses, such results only suggest that there is a relationship between the outcome of interest and the intervention, but the direction of that relationship is not determined. The results are not directly attributable to the receipt of the intervention. Importantly, evaluation designs that follow only intervention recipients over time are classified as correlational analysis without controls because there is no adjustment for sample selection or bias in these analyses. A summary of each type of study is presented in Figure A.1.

Figure A.1. Evidence Quality Types

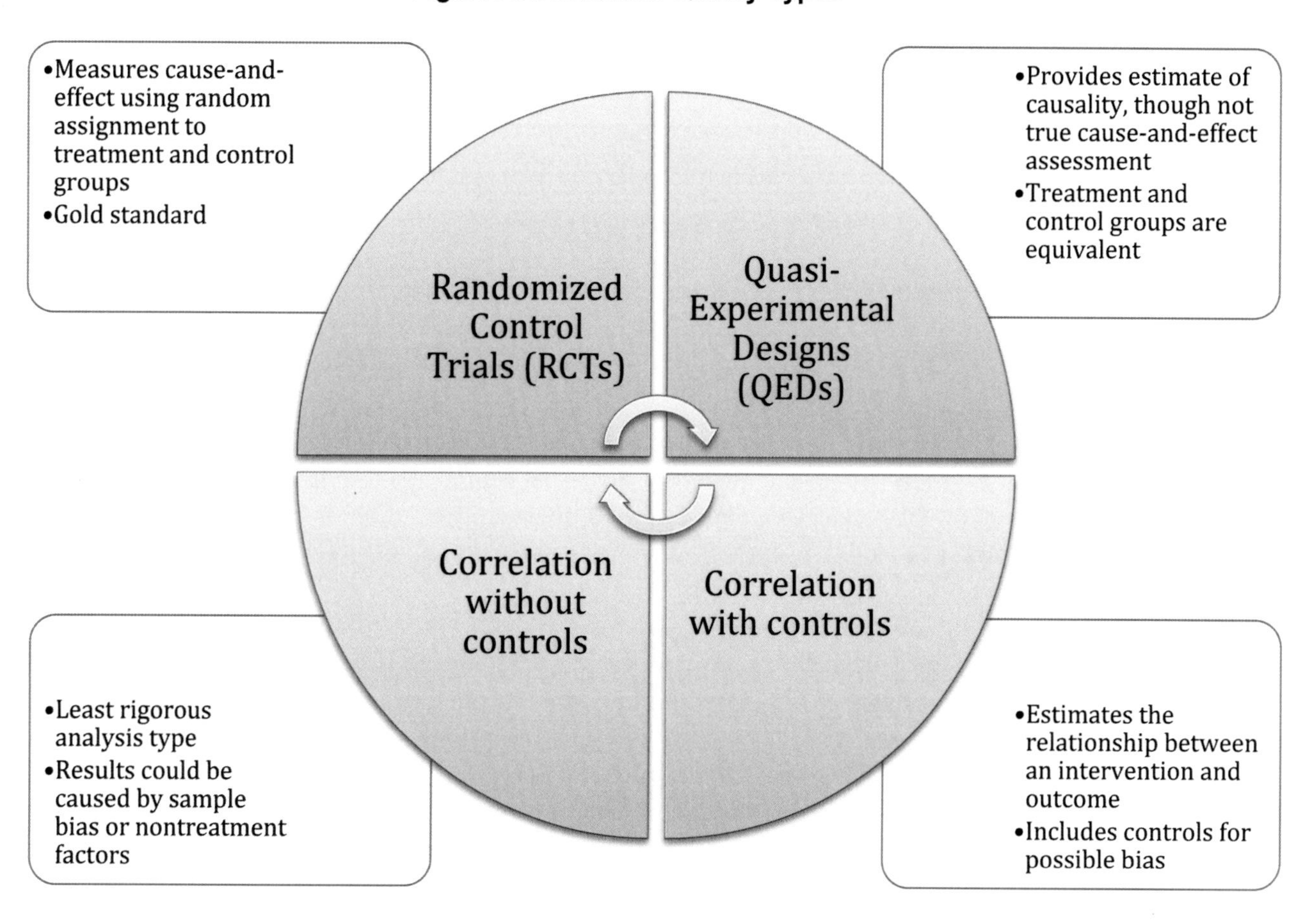

A common practice when synthesizing research findings of multiple studies on a single intervention is to look at the "totality" or body of evidence on that intervention. Multiple studies on an intervention may produce mixed evidence about the intervention's ability to influence cohesion. In our review of interventions, we reviewed the totality of evidence rather than focusing on just those findings with positive results. In the event of conflicting results, we present the findings from the most rigorous evaluation design first. For example, a study using a QED approach would be presented prior to results from a correlational analysis.

Cohesion Interventions

The majority of research available on cohesion interventions (i.e., 19 of the 29 reviewed studies) employed correlational studies without controls for sample selection. Only six studies employed a QED or RCT approach. Thus, there is limited causal evidence to support or help select interventions that build cohesion in small groups.

Intervention Types

In this section, we summarize each type of intervention identified in the literature, including a brief statement of evidence quality (see also Table 2.2). Across the 29 studies, we identified 14 types of cohesion interventions. In our review of the intervention literature, we identified interventions that can be classified by a named approach (e.g., collaborative story building and telling [CSBT]) as well as interventions that were similar in nature but that do not have a branded label (e.g., ropes courses). Where appropriate, we have grouped intervention studies together based on the type of intervention analyzed in the study. The following are the 14 intervention types:

- **CSBT.** CSBT is a practice by which a group of people share personal experiences and then work together to construct a group narrative, or story. For example, participants might be given a topic (e.g., acclimating to military life) and asked to discuss the topic within their group to create a cohesive story that represents the group's experience and perspective on the topic. We identified one intervention study on CSBT (Treadwell et al., 2011). Using a correlation without controls design, the authors found a positive association of CSBT with cohesion.
- **Collective movement.** Collective movement refers to the process by which a group of people either conceptualize moving together in synchrony or practice moving together in synchrony. An example of a collective movement intervention is a group fitness class in which everyone is moving synchronously (i.e., doing the same movement at the same time). In Göritz and Rennung, 2019, participants of the synchronous collective movement group were compared with individuals who participated in a circuit-style workout (i.e., participants were completing different exercises at the same time). Göritz and Rennung, 2019, and Wilson et al., 2018, both used a correlation without controls research design and found positive associations of collective movement with group cohesion.
- **Goal-setting.** Team goal-setting refers to the process by which a group of people, such as a sports team, identifies a set of goals that it wants to achieve over the course of a defined time frame (e.g., a sports season). Throughout the defined time frame, the group reviews progress toward its identified goals and makes adjustments as necessary. Two studies in our review investigated this type of intervention. Senécal, Loughead, and Bloom, 2008, deployed an RCT to identify the causal effect of goal-setting on group cohesion and found the intervention has a positive effect on group cohesion. The second study (Durdubas, Martin, and Koruc, 2020) found conflicting results or found that goal-setting was negatively associated with cohesion; however, the study relied on a correlational analysis with controls. More weight should be given to the positive, causal evidence from the RCT, given the discrepancy in the rigor of analytic approaches.

- **Group exercise.** Group exercise refers to the completion of a set or subset of exercises performed as a group instead of individually. One study in our review looked at group exercises and concentrated on the warm-up and cool-down sessions (Annessi, 1999). All participants engaged in warm-up exercises, an assigned workout program, and a cool-down session. The treatment participants were assigned a leader who would guide the group to perform the warm-up and cool-down exercises as a group. The researchers found increased cohesion after five weeks but no difference in cohesion after 15 weeks. The study employed a correlation without controls design.
- **Group problem-solving.** Group problem-solving refers to gathering a group of people who work together to collectively solve challenges that the group is facing. A trained facilitator leads several sessions, during which challenges to group functioning are identified and solutions are collaboratively generated by participants. In some instances, feedback from previous sessions informs the content of future sessions. Two studies investigated this approach, both working with nursing units at a teaching hospital (Barrett et al., 2009; DiMeglio et al., 2005). Both studies employed a correlational study without controls research design, and both studies found positive associations between group problem-solving and group cohesion.
- **Journal-sharing.** Journal-sharing refers to a process by which members of a small group write journal entries describing events in their lives. Participants are asked to share their entries. We found two studies that analyzed journal-sharing and found some variation in the structure of the sharing component. For example, in one study (Oh et al., 2018), some participants were asked to share their entries with just a facilitator, while others were asked to share their entries with other members of the group. In a second study (Steen, Vasserman-Stokes, and Vannatta, 2014), all participants were directed to share their entries with the group. Oh et al., 2018, used a correlation with control design, while Steen, Vasserman-Stokes, and Vannatta, 2014, used a correlation without controls design. In both cases, the intervention was negatively associated with the development of group cohesion.
- **Leader training.** Leader training refers to training the leader of a group on how to extract better performance from the group and improve group functioning, though the exact content of the leadership training can vary. We found two studies that looked at leadership training. The first (McLaren, Eys, and Murray, 2015) trained youth soccer coaches on such concepts as teamwork, cooperation, reinforcement and encouragement, positive group norms, and assigning set roles to players. The other study (Arthur and Hardy, 2014) trained officers who oversaw the training of military recruits in the United Kingdom. The officers were trained on motivation, vision, support, and coaching skills. McLaren, Eys, and Murray, 2015, employed a correlation with control design, and Arthur and Hardy, 2014, employed a correlation without control design. Both found a positive association between leadership training and group cohesion. McLaren, Eys, and Murray, 2015, saw an increase in task and social cohesion. By contrast, Arthur and Hardy, 2014, saw no change in cohesion in the intervention group, though the nontreatment group experienced decreased cohesion, suggesting that the training may have mitigated a decline in cohesion over time.
- **Mindfulness training.** Mindfulness training refers to training participants on such practices as controlled breathing, emotion regulation, and an attitude that is oriented toward curiosity, openness, and acceptance. One study in our review (Cleirigh and

Greaney, 2015) focused on mindfulness training and had groups of Irish college students listen to an audio recording that introduced breathing exercises. Students then applied mindfulness techniques to emotional experiences. The researchers employed a correlational design with controls and found that mindfulness training was associated with increased cohesion.

- **Personal disclosure mutual sharing (PDMS).** PDMS is the sharing of personal stories with a group, such that members become emotionally engaged. One study in our review (Windsor, Barker, and McCarthy, 2011) applied this technique with members of a United Kingdom soccer team. Members of the team took turns making speeches that centered on personal stories. Using a correlational design without controls, the researchers found no association between PDMS and cohesion.
- **Resilience training.** Resilience training refers to an intervention in military populations aimed at increasing the ability to cope and process stressful events. One study in our review (Adler et al., 2015) provided resilience training to U.S. soldiers during Basic Combat Training. The training focused on normalizing common reactions to stress and encouraging optimism. It also taught techniques to manage those reactions, including helping members of the group support each other. Finally, the training distinguished stressors that are under a person's control and those that are not. The study, a, RCT, found that the resilience training led to a decrease in cohesion when compared with the control group, which received a military history course.
- **Ropes course.** Ropes courses represent a class of common interventions in the cohesion literature. Ropes courses are a set of challenges that require members of the group to cooperate to complete successfully. The nature of the challenges can vary. For example, one study (Eatough, Chang, and Hall, 2015) studied a high-ropes and zipline course used in corporate retreats. Another study (Clem, Smith, and Richards, 2012) tested physical and intellectual challenges, such as group juggling, trust falls, and solving of puzzles and riddles. We found five studies, all employing correlation without control designs, that investigated variations of this approach. The results were mixed. Two studies found no relationship of ropes courses with cohesion (Ibbetson and Newell, 1999; Meyer, 2000), three studies found a positive relationship of engaging in a ropes course on cohesion (Birx, Lasala, and Wagstaff, 2011; Clem, Smith, and Richards, 2012; Eatough, Chang, and Hall, 2015), though Birx , Lasala, and Wagstaff, 2011, observed that the initial positive relationship disappeared after one academic semester.
- **Songwriting.** Songwriting refers to an intervention in which a group of people cooperate to compose a song, including music and lyrics. The one study in our review had a clinical therapist facilitate a group of clinically depressed patients through the process. The group was provided musical instruments, posters on which to write lyrics, and a recorder with which to record the final song. The author employed a correlational design without controls and found no relationship with cohesion (Cordobés, 1997).
- **Team-building protocol, Carron and Spink.** Team-building protocols refer to a set of activities on which team leaders are trained and that, when applied by the leader, improve team functioning. Carron and Spink, 1993, developed a team-building protocol that has been studied in the literature. Their approach integrates five critical activities, though the activities that occur within each step are tailored to the specific context in which the protocol is being applied. Those steps begin with training the leader of the group (e.g., an exercise group instructor) on the benefits of group cohesion. Leaders are then trained to

build cohesion by emphasizing group norms, individual positions within the group, distinctiveness of the group, individual sacrifices for the group, and social interactions among group members. Seven studies in our review analyzed the Carron and Spink team-building approach. Three of the studies (Bruner and Spink, 2010; Carron and Spink, 1993; Spink and Carron, 1993) were well-executed RCTs, and all found positive effects of the team-building protocol on cohesion. One study employed a QED (Prapavessis, Carron, and Spink, 1996) and found negative effects of the protocol on cohesion. The final three studies (Forrest and Bruner, 2017; Irwin et al., 2016; Watson, Martin Ginis, and Spink, 2004) employed correlations without controls and found positive relationships between the protocol and cohesion. The preponderance of evidence, including the most rigorous RCT designs, suggest that the use of Carron and Spink's team-building protocol increases cohesion.

- **Team-building activities, general.** Whereas the aforementioned team-building protocol revolved around a defined set of activities by Carron and Spink, 1993, this intervention (Hughes, Rosenbach, and Clover, 1983) created a protocol that was an amalgam of many approaches found in the literature of the time. The protocol included overcoming challenges together, identifying how the group performs well and where there are challenges, conflict resolution, goal-setting, and sharing personal perceptions with other members of the group. The protocol was studied on a sample of Air Force Academy Cadet squadrons during a retreat. The researchers employed correlational analyses without controls and found a positive association between the use of the protocol and cohesion.

Table A.2 summarizes our cohesion intervention evidence.

Table A.2. Cohesion Intervention Evidence, Including Intervention Type, Supporting Manuscripts, Findings, Analytic Approach, and Confidence in Findings

Intervention Type	Manuscript	Findings	Analytic Approach	Confidence in Findings
CSBT	Treadwell et al., 2011	+	Correlation, no controls	Low
Collective movement	Göritz and Rennung, 2019	+	Correlation, no controls	Low
	Wilson et al., 2018	+	Correlation, no controls	Low
Goal-setting	Senécal, Loughead, and Bloom, 2008	+	RCT	High
	Durdubas, Martin, and Koruc, 2020	–	Correlation with controls	Low
Group exercise	Annessi, 1999	+	Correlation, no controls	Low
Group problem-solving	Barrett et al., 2009	+	Correlation, no controls	Low
	DiMeglio et al., 2005	+	Correlation, no controls	Low
Journal-sharing	Oh et al., 2018	–	Correlation with controls	Low
	Steen, Vasserman-Stokes, and Vannatta, 2014	–	Correlation, no controls	Low
Leader training	McLaren, Eys, and Murray, 2015	+	Correlation with controls	Low
	Arthur and Hardy, 2014	+	Correlation, no controls	Low

Intervention Type	Manuscript	Findings	Analytic Approach	Confidence in Findings
Mindfulness training	Cleirigh and Greaney, 2015	+	Correlation with controls	Low
PDMS	Windsor, Barker, and McCarthy, 2011	0	Correlation, no controls	Low
Resilience training	Adler et al., 2015	–	RCT	High
Ropes course	Birx, Lasala, and Wagstaff, 2011	+	Correlation, no controls	Low
	Clem, Smith, and Richards, 2012	+	Correlation, no controls	Low
	Eatough, Chang, and Hall, 2015	+	Correlation, no controls	Low
	Ibbetson and Newell, 1999	0	Correlation, no controls	Low
	Meyer, 2000	0	Correlation, no controls	Low
Songwriting	Cordobés, 1997	0	Correlation, no controls	Low
Team-building, Carron and Spink	Bruner and Spink, 2010	+	RCT	High
	Carron and Spink, 1993	+	RCT	High
	Spink and Carron, 1993	+	RCT	High
	Prapavessis, Carron, and Spink, 1996	–	QED	Moderate
	Forrest and Bruner, 2017	+	Correlation, no controls	Low
	Irwin et al., 2016	+	Correlation, no controls	Low
	Watson, Martin Ginis, and Spink, 2004	0	Correlation, no controls	Low
Team-building, generic	Hughes, Rosenbach, and Clover, 1983	+	Correlation, no controls	Low

NOTE: Manuscripts are listed in the rank order of evidence quality and then alphabetically by first author. In the findings column, a positive sign (+) indicates the study found a positive association, a zero (0) indicates no association, and a negative sign (−) indicates a negative association with cohesion. Confidence in the findings is determined by the rigor of the research designs. Correlational studies with or without controls are designated low confidence because they cannot control for all factors that could affect cohesion. QED studies are designated moderate confidence because they can estimate the causal effect on cohesion under certain assumptions. RCTs are the most rigorous research design and produce high confidence in the results.

In Table A.3, we present the crosswalk between the predictors of cohesion, cohesion intervention types, and resilience and readiness building blocks. Note that this table is similar to Table 2.1 but provides the specific intervention type from the literature in addition to a high-level description of the aspect of the intervention that mapped onto the predictor. This table highlights the fact that certain cohesion predictors were represented across multiple interventions and that many of the interventions reflected several cohesion predictors.

Table A.3. Crosswalk Between Predictors of Cohesion, Cohesion Intervention Themes, and Resilience and Readiness Building Blocks

Predictor of Cohesion	Cohesion Intervention Theme (with specific examples)	Building Block
Demographic: social characteristics	--	• Demographic characteristics (background)
Demographic: work/military	--	--
Group communication	--	--
Healthy social interaction	Opportunity to share personal experiences with the group (present in CSBT; journal-sharing; mindfulness training; PDMS; team-building, Carron and Spink; team-building, generic)	• Social support (peer/squadron)
Individual subjective cognitions	--	• Sense of belonging (individual) • Sense of community (community)
Leadership	Training for leaders on teamwork, developing positive group norms, social interactions within group (present in goal-setting; leader training; team-building, Carron and Spink)	• Peer/squadron values (peer/squadron) • Community/Air Force values (community)
Mental health	--	• Mental and behavioral health (individual)
Personality	--	• Social and emotional competencies (individual) • Positivity (individual)
Group culture	Teaching team to support each other, establishing distinctiveness of the group (present in resilience training; team-building, Carron and Spink)	• Sense of community (community) • Social support (peer/squadron) • Social network (peer/squadron) • Social capital (community) • Peer group/squadron values (peer/squadron) • Community/Air Force values (community)
Group dynamics	Developing skills for group to work collectively to set goals, solve problems (present in goal-setting; group problem-solving; mindfulness training; PDMS; team-building, Carron and Spink; team-building, generic)	• Social network (peer/squadron) • Social support (peer/squadron) • Social capital (community) • Sense of community (community)
Shared culture	--	• Sense of community (community) • Peer group/squadron values (peer/squadron) • Community/Air Force values (community)
Shared experience	Providing opportunity for group to share in an activity or event (present in mindfulness training; collective movement; CSBT; group exercise; ropes course; songwriting; team-building, Carron and Spink)	• Sense of community (community) • Social capital (community) • Access to community activities (community) • Stress and strain (background)

NOTE: Some cohesion predictors have only a partial match with the identified building blocks. A cohesion predictor might only have a direct connection with certain components of a given building block, and not all aspects of a cohesion predictor were necessarily represented in the building blocks model.

Intervention Evidence Summary

There is some evidence to support the use of interventions for increasing or improving group cohesion. For example, there is evidence to support the use of Carron and Spink's team-building protocol, whereas the evidence suggests that resilience training may not be an appropriate intervention to develop cohesion, though it may be useful for other outcomes. However, we caution readers to understand the limitations of the literature on cohesion interventions.

First, the literature lacks a well-informed theory to identify the mechanisms of interventions that drive the development of group cohesion. Using the collective movement intervention as an example, we do not know whether it is the physical activity that participants engage in or the simple act of sharing time and space with others that influences cohesion. For organizations looking to develop new cohesion interventions, the literature provides limited guidance of what components those interventions should include. Second, the majority of evaluations of group cohesion interventions lack rigor and use distinct populations (e.g., individuals with depressive disorders, university counseling students) that may not respond to an intervention the same way members of the general public or military personnel would. They also largely use data from one-time snapshots of cohesion. That is, the evaluations assess cohesion outcomes at one point in time, typically right after the delivery of the intervention, and they do not examine long-term outcomes or the outcomes of interventions that are sustained over long periods (e.g., one hour of weekly team-building for an entire year). Combined with the lack of rigor in study design, this limitation makes it difficult to confidently ascribe differences in cohesion to the intervention, as opposed to other characteristics of the population who chose to engage in the intervention. Third, our search for intervention evidence turned up relatively few evaluations of cohesion interventions. It is possible that future evaluations or replications of the studies included in our analysis might provide different results if tested in more-recent work settings, with different populations of participants, or with more-rigorous statistical techniques and data. More research is needed to help guide the design and selection of future group cohesion interventions.

Appendix B. Qualitative Evaluation Methods: Process Evaluation

In this appendix, we review the analytic approach used in the process evaluation piece of the study (see Chapter 3). The process evaluation is designed to assess whether UNITE is being implemented as designed and focuses on the resources and inputs, activities, and outputs. This analysis largely leverages qualitative data provided by C3s, unit commanders and POCs, and UNITE participants. Data come from select in-depth interviews with C3s, an email survey of all C3s, AARs, and responses to an open-ended item on the participant survey.

C3 Interviews

Methods

In fall 2019 (September through December), we conducted interviews with C3s who were administering UNITE at various locations across the Air Force.[30] C3s were selected to ensure that the interview sample was representative of the total C3 population according to the following installation-level characteristics:

- MAJCOM
- number of personnel assigned to installation
- population density of the local area
- Joint or Air Force–only installation
- CONUS or OCONUS.

We recruited C3s via email, and 22 of the 24 C3s contacted agreed to participate.[31] The 22 interviewees constituted 31 percent of the total C3 population, and the sample was similar to the population in all of the installation-level characteristics.

Interviews were conducted via telephone and ranged in length from 45 to 60 minutes. We used a semistructured interview protocol, meaning that our interview protocol set forth opening questions and clear instructions, but we had discretion to pursue potentially fruitful lines of inquiry as they emerged. Semistructured interviews permit the interviewer-participant dialogue to flow as necessary to explore issues thoroughly. The semistructured nature of the interviews

[30] Before conducting the interviews, the study team satisfied all human subjects protection and data collection requirements for both the RAND Corporation and the Air Force. The interviews were an Air Force–approved data collection effort, Survey Control Number AF19-087A1Sb.

[31] The two who declined to participate were soft refusals; they responded to the initial email but not to subsequent efforts to schedule an interview.

also meant that some of the follow-up questions we posed to interviewees varied.[32] Interview topics included training and other resources available to C3s, UNITE dissemination strategies, processes related to working with squadrons to plan a UNITE activity, barriers to implementing the program, and recommendations for improvement. The interview protocol used to guide the interviews is provided at the end of this appendix.

The interviews were documented by a dedicated notetaker and analyzed using a computer-assisted qualitative data analysis procedure referred to as *coding*. Codes are labels used to organize qualitative data by topic and other characteristics (Miles and Huberman, 1994). We coded the interviews via QSR NVivo 12, a software package that permits its users to review, categorize, and analyze qualitative data, such as text, visual images, and audio recordings. After researchers assign codes to passages of text, they can later retrieve passages of similarly coded text within and across source documents, such as interview notes.

The study team developed a coding *tree*: a set of labels for assigning units of meaning to information compiled during a study, which became the basis for a codebook the team developed to clarify how the codes would be applied (DeCuir-Gunby, Marshall, and McCulloch, 2011). The codebook contained code names, definitions, inclusion and exclusion rules, and examples of interview passages that corresponded to each code. We employed a "structural" coding approach for this study; codes were based on our study objectives and interview questions and were intended to help us identify themes (Saldaña, 2015). Several members of the study team worked to develop and apply parent codes (i.e., the highest-level codes) to the interviews. This process entailed several rounds of double-coding two or three interviews, running reliability tests to check for intercoder agreement, revising the codebook, and repeating the double-coding until intercoder agreement was sufficiently high to proceed with a single coder applying the full set of parent codes to their portion of the 22 interviews. After the parent-level coding was completed, the study team reconvened to develop *child codes*: a set of additional codes intended to parse out parent codes into discrete themes. The codebook was revised to include the new codes, and all study team members then applied the new child codes to the parent codes. During this second round of coding, a single researcher was responsible for applying a subset of child codes to all the interviews in close coordination with the task lead, which meant no interrater reliability checks were needed.

After the coding was complete, we generated coding reports that enabled us to review all the passages tagged with a specific code together. For example, we analyzed the coding reports related to different barriers and challenges described by the C3s. We also used the coding reports to examine the different roles that C3s reported serving in and compared them with official role guidance provided in the Standard Core Personnel Document and January 2019 UNITE Initiative CONOPS. Members of the study team reviewed interrelated coding reports and drafted memos

---

[32] For more information regarding the use of semistructured interviews—in particular, for expert or elite interviewing—see Aberbach and Rockman, 2002, and DiCicco-Bloom and Crabtree, 2006.

that included a summary of themes, the evidence supporting each theme (e.g., prevalence, data richness, extent of contradictory examples), and exemplar quotes. The distillation of raw coding reports into these memos facilitated the development of this report.

Our C3 interview protocol is provided in Box B.1.

Box B.1. C3 Interview Protocol

1. Please describe your experience with or knowledge of Force Support Squadron (FSS) and/or Morale, Welfare, and Recreation (MWR) programs and activities prior to starting the Community Cohesion Coordinator (C3) position?
2. What training, if any, did you receive for the C3 position? How would you describe that training (probe: Was it useful? Sufficient? Timely?)? What additional training do you think would be helpful for new or current C3s?
3. As of today, approximately how many UNITE programs have you coordinated (start-to-finish)? Approximately how many commanders are you currently working with to coordinate their UNITE program?

We want to ask some questions about the process of working with a unit to coordinate an event.

Initiation of an event:

4. How do unit commanders learn about UNITE?
5. What level of leadership typically initiates a request for activities?
6. How do they connect with you to participate in UNITE?

Planning/Coordination:

7. Once you have been in touch with a commander to plan a UNITE event, what are the next steps?
8. Who identifies the program or activity? (Probe: C3, commander, some combination)
9. What types of forms do you need to complete as part of the request? (Probe: SharePoint form, funding request)

Execution:

10. What role do you have in the day-of execution of the event? (Probes: Attend event? Available to address problems that arise?)

Follow-Up:

11. What is your role after an event has taken place?
12. What type of information do you collect from the group about their experience?

13. Where do you get information and resources for your job? How do you find available programs/events?
    Prompt: For example, some C3s have knowledge of activities in the community, partnerships in the local community, and/or interaction with other C3s
14. Based on your experience, what are the benefits of this initiative for units?
15. What are the obstacles or challenges that you have faced to implementing UNITE?
    Prompt: For example, some C3s don't know about activities to use, how much outreach to do, don't know enough about the overarching goal of UNITE, or have found that funding is difficult to acquire
16. What additional information or support would make your job easier/improve your ability to fulfill your C3 responsibilities?
    Prompt: Some examples include additional information about unit cohesion, introductions to unit leadership, additional funding.
17. What else should we know? What should we be asking Commanders about UNITE?

C3 Email Survey

Methods

In addition to the in-depth C3 interviews described in the previous section, all C3s received a short email survey. The survey consisted of six questions:

- How often do commanders come to you with a specific UNITE activity in mind?
- How often do you give commanders suggestions for UNITE activities?
- When you give suggestions to commanders, how do you identify potential activities?
- What is the single biggest obstacle you face as a C3?
- What single thing would you change about UNITE to make it more effective or efficient?
- Is there anything else you would like to share about UNITE?

Survey data was collected between February and March 2020. At the time, there were 73 C3s; of those, we received responses from 40, for an overall response rate of 55 percent. One C3 had moved on to a new position, and we were unable to obtain the contact info for her replacement.

An initial round of coding and summarization was completed by the senior project lead using Excel. Another team member reviewed the original coding and summary. The results do not include raw numbers or percentages of C3s who endorsed specific ideas; however, they do provide example quotes. Although the response rate was exceptionally high for a survey, we cannot assert that there is no bias associated with who did and did not respond.

After Action Reports

Methods

After each UNITE activity, C3 requested responses to a set of five questions from each unit POC or commander who had held an event. These questions included the following:

- What FSS or off-site establishment(s) did you partner with for this event?
- What went well for this event?
- What areas needed improvements for this event.
- Would you do this event again? Why or why not?
- What lessons were learned and what recommendations do you have for future squadron events?

The analysis included a random selection of open-ended response to the questions about what went well, what improvements are needed, what lessons were learned, and recommendations for future events. We did not examine the list of specific vendors, nor did we examine data from the question about repeating events. A preliminary review of the data revealed that very few respondents (i.e., less than one-half of a percent) indicated that they would not do the event again. The preliminary review also suggested that there could be useful information in response to the “why” part of the question but that the information likely was largely redundant with other questions in the AARs. Finally, because responses were mixed between reasons why POCs would do a generic UNITE event again and a specific activity again, we chose not to review them in detail.

Selection of AARs was achieved by selecting up to four events offered by each installation using a random number generator. For installations that hosted four or fewer events, all events were included. The universe of cases was limited to events in the period overlapping with the participant survey (July 27, 2019 to December 6, 2019) and those listed as complete in the AARs. The total sample was 278 AARs.

Coding involved four members of the team. The senior project lead reviewed 150 to 200 AARs to develop initial parent codes. The junior project lead reviewed parent codes and a different set of 150 to 200 events to determine applicability and coverage of responses. We identified codes with the intent of reaching saturation, such that codes were available to cover all the responses. Several subthemes were identified to reflect distinct categories of information within each code, with examples of each provided to coders. The project leads met with two junior project staff to discuss codes, and each junior project staff member coded roughly 30 responses. The project leads provided feedback on the initial coding exercise. Using that feedback, the junior project staff coded the remaining cases, which were reviewed by the senior project lead. Discrepancies were addressed by the junior project lead. The senior project lead reviewed the final codes and synthesized results, focusing on the scope of themes rather than absolute counts of responses. All coding was done in Microsoft Excel.

Three possible limitations associated with this data source are worth noting. First, some comments may be impressions by C3s, rather than direct input by POCs or unit commanders. What went well, what did not go well, and recommendations for the future of UNITE could vary greatly between these two groups, given their different vantage points of the unit. Second, some events were combined with other unit-wide events, such as a resilience day or a stand-down because of a suicide. Thus, it is not entirely clear whether comments are really about UNITE, the other event, or both. Third, the analysis used only a sample of all AARs. With more than 2,400 UNITE activities in the C3 SharePoint database provided to us that met the selection criteria, we could not code every open-ended response. Although our sampling strategy should provide us with a random sample of events, we did not capture every recommendation or lesson learned provided in the AARs.

Detailed Results

Tables B.1, B.2, and B.3 present results for the three AAR questions of interest (what went well, what areas need improvement, and lessons learned, respectively), focusing on the scope of themes found in the responses rather than absolute counts of responses. Each table provides the main themes that emerged from our analysis of the AAR data, an explanation of the theme, and example quotes (when applicable).

Table B.1. UNITE Event After Action Reports: What Went Well?

Theme	Explanation	Example Quote(s)
Everything	Refers to comments that explicitly indicated that all aspects of the event went well. It is a mutually exclusive code, in that it is not used with any other theme.	• We enjoyed the event very much! Everyone had a great time and enjoyed the time away from the clinic! Everything went very well. Everything was very well communicated. We knew when to be there and who to speak with. No issues!
UNITE event provided an opportunity for improved morale, esprit de corps, or camaraderie	Refers to comments that specifically call out UNITE activities as an opportunity to build unit morale, esprit de corps, or camaraderie. Generally, these comments reflected that the group nature of the UNITE event led to an increase in group morale or camaraderie.	• The event was a great success, it afforded the opportunity for all squadron members to build camaraderie and moral[e]. • Squadron [m]orale was increased and now at an all-time high.
UNITE event provided an opportunity for improved cohesion	Refers to comments that specifically call out UNITE activities as an opportunity to improve the cohesion of the group. The word "cohesion" did not necessarily have to appear in the comment to be included in this theme. The cohesion code was used when comments explicitly mentioned "cohesion" or "unity," when they explicitly mentioned working toward a common goal (i.e., task cohesion), or when they mentioned "bonding" or being brought together as a group by an event (i.e., social cohesion). Many of the comments in this theme also included elements that were coded in the morale, esprit de corps, and camaraderie code, which is perhaps unsurprising given the overlap between cohesion and these constructs. However, the morale, esprit de corps, and camaraderie code was used when the comments conveyed feelings that resulted from cohesion or bonding activities.	• The event was raved about by Squadron leadership and helped boost morale and cohesiveness for our entire unit. • Bringing 500 people together is difficult, but the funds from this program allowed us to put together an event that everyone WANTED to come to. It definitely inspired unity.
UNITE event provided an opportunity for team-building	Refers to comments that specifically called out UNITE activities as a means to build a team, create teamwork, or enhance working together. The concept of team-building is also related to the prior two themes (i.e., morale/camaraderie and cohesion), and specific mentions of team-building were often combined with similar constructs.	• The event went really well overall. The activities that were available allowed members to pair up into teams and discover their strengths and weaknesses. The communication and teamwork aspect was exceptional! • [UNITE] funds support allowed for our squadron to come together in a relaxed setting to discuss lessons in leadership and teamwork, as depicted in the movie, and how that can apply in the squadron on a daily basis to open the door to solve problems as a team now and in the future.

Theme	Explanation	Example Quote(s)
Opportunity for competition	Refers to comments that specifically call out UNITE activities that have an aspect of competition among participants. Many of these comments highlighted that the competitive aspect of activities (typically, healthy competition among unit members) provided opportunities for team-building.	• Some things that went well include the team building, fun and fellowship and friendly competitions within the squadron. There was an excellent time had by all and it was a smooth process.
UNITE event provided an opportunity for unit interaction and socialization	Refers to comments that specifically call out UNITE activities as an opportunity for unit members to interact, socialize, or get to know one another better. The majority of these comments were variations on a similar theme: Group activities for the entire unit provide an opportunity to get to know other unit members in a relaxed way that is not always possible within the stressful confines of day-to-day job activities.	• The ice breakers allowed members to get to know people from other sections that they were not familiar with. The bowling allowed members to break up into teams made up of people from different sections allowing teamwork between members that would not normally socialize. • Participants enjoyed the activity. Due to working a swing shift, the unit does not often get a chance to come together and have fun as a unit. The atmosphere was relaxed which is great for [unit] as their normal day-to-day activities can be stressful.
UNITE event provided an opportunity for improved resilience	Refers to comments that specifically call out UNITE activities as an opportunity to improve unit resilience. In some instances, it appeared that UNITE activities were combined with resilience pauses or other unit-wide events or trainings. This joint nature of some events was reflected in comments about what went well.	• Additionally, we conducted this event during resiliency time on a training day, which allowed airmen from different shifts to attend. This went a long way to building the cohesion that was the goal of the event.
UNITE event provided time to relax, rejuvenate, release stress, or have fun	Refers to comments that specifically call out UNITE activities as an opportunity for unit members to relax, rejuvenate, release stress, or have fun. The majority of these comments were variations on a similar theme: UNITE activities allow unit members to step back from their busy, often stressful, work environment and simply have fun and relax with their fellow unit members. Comments in this theme also overlap with those in other themes, especially morale, cohesion, teamwork, and socializing.	• Squadron members were in high spirits, very thankful, and truly enjoyed the opportunity to decompress. • The event was a great success, it afforded the opportunity for all squadron members to build camaraderie and moral[e]. It was a fun way for us to relax and promoted positive interaction between all members and all ranks.
Getting out of the office	Refers to comments that specifically call out UNITE activities as an opportunity for unit members to get out of the office and away from work responsibilities. Holding UNITE activities outside the office or shop setting was specifically mentioned by enough respondents that it warranted its own code. Although it is not possible to use the AAR data to ascertain whether events held away from work were more successful than those that were not, it is still noteworthy that the location of the event appeared as a theme in the comments.	• This was a great event for members of the squadron to go out and build unit cohesion in a non-work environment. • Our Division was able to come together, bond as a team. Getting away from our desks to communicate and express our thoughts and feelings, and not just communicate about work.

Theme	Explanation	Example Quote(s)
Funding and cost	Refers to comments that specifically referred to funding aspects of UNITE, which could include the availability of funding, the amount of funding, or rules and regulations related to use of funding. Respondents highlighted that the availability of UNITE funding allowed units to do things they ordinarily would not have been able to do. Other respondents commented on how easy it was to use the funding.	• The overwhelming reaction from the flight was positive and full of gratitude. Many members felt that it had been far too long since the flight had been able to spend an evening together at an event that did not require them to pay out of pocket. With the dinner cruise happening in the evening, it ensured that the two crews would see each other and spend some quality time with their peers and leaders without interfering with their sleep schedules which was greatly appreciated by all in attendance.
Logistics and execution	Refers to comments that specifically call out the logistics and execution of UNITE activities, which could include issues related to vendor services or location, delivery, food and beverage quality or quantity, equipment and rentals, and transportation. The majority of these comments were accolades about how the event played out, with many noting participant satisfaction with activities and food, participation and interaction by participants, and positive interaction with vendors and venues	• We were happy with how many people showed up, despite it being a holiday weekend. The service from [VENDOR] was fantastic. They opened early to accommodate our group and even threw us a tournament. They offered us a great price as well. People really enjoyed the activity itself. They loved the adrenaline rush.
Interaction with C3	Refers to comments that specifically call out interactions with C3 during the UNITE planning process. Most of these comments noted how having a C3 facilitated the planning and execution of the event.	• The whole experience was awesome and way easier to do with [C3] being so willing to assist with our every need. • [C3] was integral to the ease and flawless flow of events, which made [it] even more stress-free for me.
Specific event or vendor mentioned in relation to positive aspects	Refers to comments that call out specific events or vendors in relation to the positive aspects of the UNITE event. Comments in this section tended to be about a specific type of UNITE activity (e.g., escape rooms, bowling) or a specific vendor (e.g., the installation's club, local caterer). This made it difficult to draw any substantive conclusions for UNITE as a whole. However, the one consistent theme was a high level of satisfaction with events that included both an activity and food, such as renting outdoor recreation equipment and a picnic.	• NA

NOTE: NA = not applicable.

Table B.2. UNITE Event After Action Reports: What Areas Need Improvement?

Theme	Explanation	Example Quote(s)
Nothing	Refers to comments that do not call out a specific area of improvement. In some cases, only positive aspects are listed. It is mutually exclusive to all other themes.	• Nothing significant to report. The event was great. No issues and everything went very as planned.
Frequency of UNITE activities	Refers to comments that address areas of improvement related to the frequency of UNITE.	• One area that needs improvement is to have it more regularly like monthly squadron event.
Funding	Refers to comments that address areas of improvement related to funding aspects of UNITE. Comments in this category clustered around four aspects of funding: issues using the funding with certain vendors, having more flexibility with the use of funding, having a better understanding of UNITE funding, and having more funding.	• The only recommendation I would offer is making another payment option. Some of our funds were lost in fees because some vendors charge a fee when running credit cards. My recommendation is to offer a check option. • I wish I was able to spend the money jointly, i.e., on both event and food . . . would have been nice to feed the Squadron while we were there, however, I didn't have any "food" money left. I think as we fine tune this program, integrity is important, we need to entrust our Commanders to spend the money as they see fit . . . and stay within the parameters of the budget given for each squadron. • Additional guidance from the UNITE team on what our unit was able to do (where we can go and what could be purchased). The UNITE [PowerPoint] presentation was complicated and left our team open to interpretation.
Participation	Refers to comments that address areas of improvement related to participation of unit members in UNITE activities. This code included comments related to scheduling (e.g., during weekends, after-hours), no-shows, family member eligibility, and deconflicting of schedules. Comments in this theme clustered around three subthemes: better understanding of attendance, especially for planning purposes (e.g., food orders); scheduling an event to maximize unit attendance; and getting more people to attend. Understanding attendance, largely for the purpose of planning but also to remind unit members of an event (or to hold them accountable in terms of RSVP'ing to an event) was a common subtheme.	• I would start the financial process earlier because there were more military members that attended than I originally put on the list. • Attendance still needs work. Many sections will not be able to attend events during duty hours, while other sections are not interested in events outside of duty hours. • Try and find a time where everyone can attend at the same time versus being able to come during different shifts.

Theme	Explanation	Example Quote(s)
Planning process before event	Refers to comments that address areas of improvement related to the planning of UNITE activities and includes aspects related to paperwork, purchasing, working with vendors, and interactions and communication with C3s. These comments clustered around two central themes: starting the planning process earlier and working with vendors to settle on details before the event.	• We need to be more on the ball with planning these events and getting them on the calendar early. We didn't leave our C3 much time to do the work on her end, she went out of her way to help us. • The only area that needed to be smoothed out more was the coordination with the actual venue. There was confusion with what was able to be done within the price range.
Planning in advance	Refers to comments that specifically mention planning in advance of UNITE activities. A specific mention of planning in advance occurred in about one-quarter of the comments that related to event planning.	• Better communication and more planning time at all levels would have [led] to a smoother event. The game plan for our event changed up a number of times in the short time frame before execution and unfortunately our main POC/planner for the event was on leave the during the week leading up to the event.
Logistics and execution during or the day of the event	Refers to comments that address areas of improvement related to logistics and execution during or the day of UNITE activities. It includes comments about weather (e.g., too hot, rained), vendor problems, food-related issues (e.g., too much, too little), rental equipment issues, and obtaining DoD IDs. Many of the specific comments are not generalizable to UNITE events as a whole (e.g., "1/4lb. burgers cook really fast and also shrivel up when trying to keep them warm"), though a few themes may be relevant for C3s and unit POCs planning future UNITE events.	• We will hold this again in the cooler part of the year . . . August was too hot for bus tours of the base, since the buses' AC [air conditioning] couldn't keep up. • Food prep time coordination. With such a large order the food took longer to make then expected, forcing most of the team members to have to wait for us to bring food back to their shops later that day. • Although the need to sign in is a little cumbersome and inefficient. Maybe a way to scan CACs [Common Access Cards] as opposed to signing in?
Marketing/advertising	Refers to comments that address areas of improvement related to marketing of and communicating about UNITE activities. Specifically, these comments addressed getting the word out to the unit about the event, typically with the assumption that this would improve attendance and participation. In addition, some comments suggested that part of the marketing for a UNITE event should include information not only about what the event is and when it will occur but also why it is happening.	• A recommended improvement area would be to better promote the party for greater participation in the event. • Continue to improve on advertisement of the event and ensure personnel feel the event is worthwhile not just for free food, but also as a teammate.
Substantive aspects of activities	Refers to comments that address areas of improvement related to general substantive aspects of UNITE activities that are not activity specific. Such comments may mention linking a UNITE event back to the reason for UNITE (e.g., increasing unit cohesion), having more time for members to interact during the event, and having structured activities. Comments here clustered around making sure that airmen understand the purpose of UNITE.	• Communication to members that this event is sponsored by [UNITE] and the purpose behind it. • I also believe that we should have benefited from having someone facilitate the event, or at least in the start of the event be present to explain the events significance. • Having things relate back to why we are doing the activity.

Table B.3. UNITE Event After Action Reports: What Are the Lessons Learned and Recommendations for Future Events?

Theme	Explanation	Example Quote(s)
Venue-related issues	Refers to comments that address lessons learned and recommendations related to venue-related issues. One important subtheme that was clearly expressed in comments was the importance of leaving the workplace for a UNITE event. Other comments captured in this theme were varied, ranging from recommendations to establish a line of communication with venues, to making sure a venue can handle the size of the planned event, to rotating venues to keep airmen interested in UNITE events.	• For future events, I hope that we try to venture out to more off site locations so that we get a chance to get away from the office. • [L]eaving the squadron to host a future event will be more beneficial as staying here it's hard to stop the mission. Too many distractions whether we want them or not. Ops never stop.
Vendor-related issues	Refers to comments that address lessons learned and recommendations related to vendor-related issues. Communication and planning in advance, especially for unexpected or last-minute changes, were the primary subthemes.	• Only recommendation is to keep communicating with UNITE POC and food or activity vendor leading up to the event. In our particular case the credit card used to pay for the food did not go through properly, I notified the POC and she immediately corrected the issue so there wasn't a delay getting our food. • Having a solid plan far in advance and communication with whatever contractors you are partnering with and [squadron] members.
Activity-related issues	Refers to comments that address general lessons learned and recommendations related to activity-related issues. It does not include lessons learned or recommendations about specific activities (e.g., identifying new or novel activities). The most common subtheme among comments was the importance of customizing UNITE activities to the individuals in each unit.	• Next time I think we would allow the individual flights the opportunity to schedule and plan their own smaller events. • Main ingredient to successfully using the UNITE funds is good planning and selecting an activity that is best suited for the people in your unit. • Future recommendations would be to ask people to please provide more input on what they would want to see or do for the next event to more people interested.
Planning process before event	Refers to comments that address lessons learned and recommendations related to the planning process. Three subthemes stood out: planning in advance, getting an accurate head count of attendees, and having a backup plan in place.	• I think planning further in advance would have opened the opportunity to having other options available, but the planning and coordination started later due to Change of Command and adding the Commander's Call into the event. • One of our lessons learned was that we need to over-estimate that attendance; then, if the final attendees are less than expected, it is easier to take games away than think of them on the spot. • We need to identify at minimum two alternate POC's . . . When I was out of office for a period of time it caused delayed communication between myself and the C3.

Theme	Explanation	Example Quote(s)
Obtaining DoD IDs	Refers to comments that address lessons learned and recommendations related to collecting DoD IDs. Most recommended an easier way to collect attendance (e.g., DoD IDs).	• [M]aking an easier way to obtain DOD ID numbers from members without making it seem like the event is "mandatory fun." By obtaining them it makes someone to commit to the event far out due to the process of approval. So that aspect makes it difficult to accurately prep for the numbers you'll have.
Attendance and participation	Refers to comments that address lessons learned and recommendations related to attendance and includes comments about increasing participation rates, planning for more or fewer attendees than expected, and advertising and communicating about events. Comments in this section largely offered respondents' recommendations on how to improve attendance, and those suggestions were quite varied.	• I would just double check the schedule to make sure everyone is able to participate on the day the event is scheduled. • Getting people to participate is always going to be difficult, but moving forward, if we are able to get word out about the event a couple weeks earlier I think that we would be able to get people excited and raise a little more awareness. • I believe that more time needs to be put into planning and notification of the event so that we might get a higher participation rate.
Frequency of events	Refers to comments that address lessons learned and recommendations related to the frequency of UNITE events. Although comments about this theme were relatively infrequent, they all shared the same sentiment: Hold more UNITE events.	• Just to have these events more often. People always enjoy events like this. It brings people together. • Do this type of activity more often so that it will help boost morale and to get to know fellow co-workers.
Funding	Refers to comments that address lessons learned and recommendations related to funding issues and includes comments about where UNITE money can be used, how to use the funding, and limitations of the current spending process. Subthemes varied from comments about understanding when UNITE funds can and cannot be used to comments about how to stretch UNITE dollars farther.	• Don't be afraid to incorporate a small out of pocket cost in order to attend worthwhile events that are more than the UNITE Funds will pay for. It's still a great discount. • Recommend future squadron events are set up on a regular basis, even if funding via the UNITE Program are already utilized, we can still build on our team via low cost potluck style events.
Substantive aspects of activities	Refers to comments that address areas of improvement related to general substantive aspects of UNITE activities that are not activity specific. Such comments might mention linking a UNITE event back to the reason for UNITE (e.g., increasing unit cohesion), having more time for members to interact during the event, and having structured activities. A wide variety of topics were covered in this theme, but two subthemes in particular received more attention: recommendations about the size of events and making sure all attendees understand the purpose of the event. Smaller events were perceived as more conducive to achieving the goals of UNITE (specifically, improving group cohesion).	• I think small group morale events are much more conducive to building comradery. • Think smaller groups definitely can benefit from UNITE funds more easily. Recommend flight sized groups [versus] unit as a whole. • Lessons to the facilitators were to provide the why for the events. Relate it back to work related topics and how it can help in the future. • I feel we would have benefited from having a third party facilitator at the event to explain why we were spending that time together, as well as its importance.

Open-Ended Participant Survey Data

At the end of both the initial and the follow-up UNITE participant surveys, respondents were offered an opportunity to provide any additional thoughts they had about UNITE in an open-ended text field.[33] Responding to this survey item was optional, and we received a total of 595 comments in reference to activities that occurred between July 27, 2019 and November 7, 2019.[34] After an initial review of the comments by the senior project lead and discussion with the junior project lead, we identified four major areas:

- What is the impact of the program or activity?
- What are the positive aspects of UNITE?
- What are the negative aspects of UNITE?
- Participant recommendations.

Methods

Coding involved three members of the team. The senior project lead reviewed 150 to 200 survey responses to develop initial parent codes. The junior project lead reviewed parent codes and a different set of 150 to 200 responses to determine applicability and coverage of responses. As with the AAR data, codes were developed to cover all responses. The senior project lead worked with one junior project staff member to discuss codes, and each coded roughly 50 responses. After receiving feedback on the initial coding exercise, the junior project staff member coded the remaining cases. The senior project lead reviewed all code assignments, and the junior project lead addressed any discrepancies. The senior project lead reviewed the final codes and synthesized results, focusing on the scope of themes rather than absolute counts of responses. All coding was done in Microsoft Excel. Obviously, not all survey respondents offered additional comments, and thus there is a risk of bias if respondents who offered a comment are systematically different from those who did not on some characteristic or UNITE-related experience. We examined differences in characteristics between respondents who provided substantive comments and those who did not. Those who offered substantive comments were more likely to be officers, be older, and have served longer. However, they did not differ on how they, on average, rated cohesion (overall, social, or task).

Detailed Results

We present themes for the four areas of interest identified in the open-ended participant survey data that related to the implementation of UNITE, focusing on the examples found in the

---

[33] The email surveys were an Air Force–approved data collection effort, Survey Control Number AF19-191A1.

[34] Note that the cutoff date for open-ended survey data is roughly one month before the cutoff date for the other, quantitative survey data (December 6, 2019). According to our initial review of the open-ended data, we found that we had saturated codes and opted not to include additional open-ended survey responses.

responses: perceived impact, positive aspects, negative aspects, and participant recommendations.

Impact of UNITE

The impact theme only applied to comments that conveyed that the UNITE Initiative or UNITE event had an *actual* impact. If the comment used words like "tried" or "attempted" to do something but did not indicate an actual impact, it was not be coded here. The majority of comments did not receive a code for this theme. That is, the majority of comments did not mention an actual impact caused by participation in UNITE. Nonetheless, we identified five themes that fell into the category of impact: unit morale, esprit de corps, and camaraderie; cohesion and unity; team-building; socialization, interaction, bonding, and communication with unit members; and opportunities to relax, unwind, and have fun. Note that there is a great deal of overlap in some of the key terms used across these categories (e.g., bonding, cohesion, unity), and this is reflected in many of the example quotes. Descriptions of themes and example quotes are presented in Table B.4.

Table B.4. Airmen Post-Participation Surveys: Perceived Impact of UNITE

Theme	Explanation	Example Quote(s)
Unit morale, esprit de corps, and camaraderie	Refers to comments that indicated that participation in UNITE events had a demonstrable impact on unit morale, esprit de corps, or camaraderie. Comments categorized under this code reflected the participant's belief that participation in UNITE provided a mechanism to "boost" existing unit camaraderie and/or build on it.	• I really enjoyed participating in the activity on training day, it was nice to do a group activity together outside of the work place, I felt like it boosted moral[e] and provided an opportunity to socialize with people in other sections. • The UNITE funding has done an excellent job to assist in creating downtime for our airmen to bond and build camaraderie. I sincerely appreciate the funds and the ability to build stronger teams.
Cohesion and unity	Refers to comments that indicated that participation in UNITE activities had a demonstrable impact on cohesion or unity among unit members.	• This was a great opportunity for our team to come together and reinforce teamwork, camaraderie and build group cohesion. We do this normally through Squadron PT [physical training], morale boosting events, and other means, but this was a nice addition for the team. • The event I participated in was awesome. I feel like it got the job done with helping out with unit cohesion. Members in my unit do not typically hang out outside of work. This provided a good opportunity for that. • It was a well-needed bonding activity for the squadron. I have heard nothing but good things about that day. It connected us as a unit and brought us together.
Team-building	Refers to comments that indicated that participation in UNITE activities provided an opportunity for units to practice team-building. In some cases, team-building was mentioned in the context of other perceived benefits, including cohesion, morale, and resiliency.[a]	• The UNITE events have been great for cohesion and resiliency outside of work. We have all enjoyed being able to socialize and team build outside of the normal work environment. • The UNITE event was very beneficial and has helped promote healthy interactions within my unit. I have seen increased morale and teamwork since then. • The activity provided was a good opportunity to start team building for the squadron. I wish to see more activities like this in the future.

Theme	Explanation	Example Quote(s)
Socialization, interaction, bonding, and communication with unit members	Refers to comments that indicated that participation in UNITE activities provided an opportunity for unit members to socialize, interact, bond, or communicate with other unit members. Often, these codes reflected participants' views that UNITE provided a way to interact with other unit members, sometimes with members they would not ordinarily interact with.	• I believe this event was a success, it allowed our unit to build [rapport] with one another that you don't typically see due to working in different buildings and different hours. It was a fun to have competition and come together for some food and conversation. • This was a very successful event. You could tell that everyone in the squadron was having a good time. It was a great opportunity for our unit to socialize together outside of work. Please keep this program active! • I think people see things as "mandatory fun" and automatically have negative thoughts towards it, but enough "mandatory fun" creates socialization and gets people to know each other, find common interests, and ultimately enjoy doing things together.
Opportunities to relax, unwind, and have fun	Refers to comments that indicated that participation in UNITE activities provided an opportunity for unit members to relax, unwind, or have fun. Like many of the comments above, when participants mentioned that UNITE activities provided an opportunity for unit members to relax, unwind, or have fun, it was often combined with some other benefit.	• Absolutely wonderful program that assisted with integrating our new members and allowing for so much needed relaxation/bonding time. Thank You. • This event was an amazing, fun way for the members to laugh, get to know each other and relax. Thank you for allowing us to have this team building opportunity. Absolutely the best event EVER!

[a] Note that we explored whether resilience should be its own code, but relatively few (about ten) of the comments specifically mentioned that resilience was a benefit associated with UNITE participation.

Positive Aspects of UNITE

Codes in the positive aspects of UNITE theme ranged from the very specific (e.g., enjoyment of a particular activity, such as bowling or an escape room) to the very general (e.g., gratitude for UNITE). Comments that were very specific are not reviewed here because they are not generalizable to UNITE overall. Our analysis identified four codes that fell into the category of positive aspects of UNITE: logistics; funding; socialization, interaction, bonding, and communication with unit members; and opportunities to relax, unwind, or have fun. Because two of these codes (socialization, interaction, bonding, and communication with unit members; opportunities to relax, unwind, or have fun) overlapped substantially with themes represented in the "program impact" theme (described in the previous section), we focus this section on logistics and funding-related comments. Descriptions of themes and example quotes are presented in Table B.5.

Table B.5. Airmen Post-Participation Surveys: Positive Aspects of UNITE

Theme	Explanation	Example Quote(s)
Logistics	Refers to comments that indicated that the actual logistical aspects of the UNITE event went well and included comments about vendors or venues, C3 assistance, or food. The vast majority of these comments were not generalizable to UNITE as a whole, though when the logistical aspects of an activity went well, participants clearly noticed.	• NA
Funding	Refers to comments that indicated that funding was a positive aspect of UNITE and included comments about the ease of using funding or about reducing cost for airmen. The majority of comments with this code acknowledged how UNITE funding alleviated pressure on units to pay for much-needed morale events.	• UNITE is a great program that allows a squadron to participate in an activity and not having to either "pass the hat" or look to the booster club for funding to do so is very helpful and encourages participation. • I greatly appreciate the UNITE funds, especially the ability to do things together without first having to figure out how to raise money during our already tapped-out days of constant high ops tempo. • Was unaware of UNITE until recently. I really liked the ability to alleviate a financial obligation on the member for events often considered "forced fun." We do have a responsibility to develop unit cohesion, and many times this is done away from work.

NOTE: NA = not applicable.

Negative Aspects of UNITE

Although respondents were largely positive about UNITE, some participants offered comments that were less than complimentary. As with the positive comments offered by participants, negative comments ranged from the very specific (e.g., dissatisfaction with specific activities) to the more general (e.g., lack of awareness about the reasons for UNITE). Specific comments are, again, omitted here because they are not generalizable to UNITE overall. Our analysis identified six codes that fell into the category of negative aspects of UNITE: logistics, funding, required participation, low participation, the substantive nature of UNITE activities, and other nonspecific negative comments that are not covered in any of the other codes for this theme. Table B.6 presents a description of each theme and example quotes.

Table B.6. Airmen Post-Participation Surveys: Negative Aspects of UNITE

Theme	Explanation	Example Quote(s)
Logistics	Refers to comments that indicated that the actual logistical aspects of the UNITE event did not go well and includes comments about vendors or venues, food, the weather, and timing. Comments here covered complaints about execution of the event, not enough food, bad weather (e.g., too hot), and scheduling of the event.	• It was a waste of a day. Poorly organized with unrealistic time tables, going over group times and not get time for small groups. Nothing about this day was "relaxed." • The amount of food provided (or funding allowed for food) was not adequate. Several members of the unit, including myself, mentioned that they would need to go get additional food for lunch after the UNITE event because they did not get enough. • Event took place on a down-day, which I consider outside of duty hours.
Funding	Refers to negative comments about features or requirements related to UNITE funding. Comments largely clustered around three subthemes: limitations on use of funds, difficulty obtaining funding, and the amount of funding.	• Process to request and use UNITE funds is cumbersome. Funds/funded activity are not worth the planning effort and admin tail for sign-ups, sign-ins, photos, and after action report. • I suspect that UNITE funds would be utilized more if the paperwork/payment process was more [straightforward]. • The UNITE program is great, raising the spending limit would only make the program better. The quality of events is slightly controlled by the spending limit. Quality events lead to quality memories and time spent. Give more freedom to the planner.
Required participation	Refers to comments that indicated UNITE activities were "mandatory fun" and/or took valuable time away from work to attend. Generally, some participants felt that taking time outside of work to participate in a UNITE event prevented them from doing their job and led to more, not less, stress when they returned to work.	• I would rather have spent time doing work. This was yet another unit event that took time out of our mission accomplishment. This caused me more stress because I had to work over the weekend to make up for missed time. This took time away from family. • To be completely honest, I would rather have stayed at work to take care of things rather than be forced to awkwardly interact with our commander and some of the other leadership team.
Low participation	Refers to comments that were related to other negative aspects of participation, including comments about low turnout, eligibility of family members, or timing of the event with respect to unit members' schedules.	• I really wish family members would be included in the funding for these events. Our families have long been considered part of our unit and to organize an event out of the office and not consider them is not something we are used to. • While it was a good idea to have the unit get together, most people left the area after eating. It did not seem like it was really a time to get together to relax because a lot of people went back to work after. • Because of the services career field, only a quarter of us were able to attend the event. These programs are great ideas, but when only 1/4 of the [squadron] can attend them, it brings down the morale of those who have to work during it. There's no easy fix.

Theme	Explanation	Example Quote(s)
Substantive nature of UNITE activities	Refers to negative comments about aspects of UNITE activities related to the substantive nature of the event itself. It also applies when a comment suggested that the event was a wasted opportunity to achieve unit cohesion. Participants' comments provided a range of subthemes, but two were more salient than others. First, some participants noted that although they enjoyed the UNITE event, the motivation for UNITE was not always obvious, either because of the lack of a clear message or problems with the messenger. Second, and related to the first subtheme, some comments reflected mixed messages about what the UNITE event was for. This seemed to be the case when the UNITE event was combined with some other event, such as a stand-down day.	• Some people miss the intent about sport days and just focus on being extremely physical. Some would just talk "junk" and put people down so they can win the game. Missed the mark that we are still one team as a whole. • Overall it was a good concept to get people to talk and open up. Facilitators were of varying effectiveness, possibly next time bring in SNCO's [senior noncommissioned officers] from other units to facilitate, younger airmen and NCO's [noncommissioned officers] are not always the best choice. • The event was about Suicide awareness, so calling it a "relaxed and fun" [environment] isn't really accurate. • The day was a success in terms of meeting new people/ building a deeper connection with people I would normally not interact with. The event was poor in the side groups hour where they divided us by rank and had us "discuss" suicide signs and prevention.
General, nonspecific comments	Refers to negative aspects of UNITE activities that were not captured in other codes. One might call this theme "miscellaneous grievances." Two subthemes emerged among these comments. First, some participants felt that UNITE as an initiative is too reactive to the problems they perceive to exist in the Air Force and reinforced that some UNITE activities were tied to a stand-down day, resulting in mixed messages about the purpose of UNITE. Second, some participants noted that they were unaware of UNITE before the event and wanted to make sure that their counterparts were aware.	• These type of events should be part of the norm, not something that needs to happen because something is driving it. • I think it is a great idea to have events like this every so often. My only issue with this event is why do we have to wait for someone to die in order for us to do this. Let's be proactive. Not Reactive. • It would have been nice to hear about this program earlier. We did not get anything on it until the CSAF [Chief of Staff Air Force] directed stand down almost a year into the program. Many others that I've talked to still do not know about it. Excellent program with little media. • I never knew that there was a program like UNITE that was available to bring the unit together in a great way. Information about the program was not advertised very well. Now that I know about the program, I have given information to my counterparts.

Participant Recommendations

The final theme we explored in the open-ended participant survey comments was recommendations. Most participants who provided comments did not include actionable recommendations for improving UNITE. Among those who did, we identified six codes that fell into the category of recommendations: logistics, funding, frequency of events, participation, and the substantive nature of UNITE activities. Perhaps not surprisingly, these themes largely mirror what we saw in the negative comments discussed earlier. Table B.7 presents a description of each theme and example quotes.

Table B.7. Airmen Post-Participation Surveys: Participant Recommendations

Theme	Explanation	Example Quote(s)
Logistics	Refers to comments that made recommendations about logistical aspects of UNITE events and that were related to the location of events and timing of activities. Some participants indicated that UNITE activities should be held during traditional duty hours or outside the installation, though these were by no means a consistent theme across comments.	• Doing unit activities DURING duty hours is awesome. But outside of duty hours it's ANNOYING. The best "moral[e] event" that we all would prefer in the Air Force is just simply to go home for the day. It's that simple. • Also, we had the hardest time getting our other UNITE event during a work day. People will be more inclined to attend if it was during a duty day. • I feel that if the event would have been held off base, it would have been more of a social event. [M]ost of us who found out last minute thought it was downtown until the day of the event.
Funding	Refers to comments that made recommendations about the funding for UNITE events. The majority of comments in this theme were related to improving the process of obtaining and using UNITE funds	• Please get the funds to be approved at a lower level. A faster response time and wider opportunity for funding would allow for more successful events. The amount of time it takes to coordinate with the agency and plan the event cause last minute execution. • [L]et them [units] use it as they see fit. Or set a percentage of what can be used per event without other paperwork being done to provide it a larger event. i.e., 30% can cover a Christmas party and the rest on other events this would help smaller units. • The UNITE program is great, raising the spending limit would only make the program better. The quality of events is slightly controlled by the spending limit. Quality events lead to quality memories and time spent. Give more freedom to the planner.
Frequency of activities	Refers to comments that made recommendations about the frequency of UNITE activities. Note that simply stating that a participant would like to have more events is not considered a recommendation. All comments in this theme suggested that the Air Force retain UNITE as an initiative, and the majority of these comments indicated that participants want a higher frequent of UNITE activities for their units.	• Events like these should become more commonplace in the squadron! Was great to get everyone together, where we wouldn't normally. Moral[e] seemed very raised during this day. • I wish that we could do something like this on a more regular basis—once every couple of months. It is important as a unit to get to know our peers and have the chance to step away from the stressors of work. It is also beneficial to the mission.

Theme	Explanation	Example Quote(s)
Participation	Refers to comments that made recommendations about participation in UNITE events. It included comments about eligibility of family members or civilians and ways to increase turnout. Note that simply stating that a participant would like to have seen more attendees at an event is not considered a recommendation. Some participants directly asked for families and civilians/contractors to be able to attend. Other comments offered suggestions for improving attendance by unit members at UNITE events.	• The UNITE Funds program should figure out how to incorporate families, however, as our unit doesn't do anything without our families. • Also, can you add in contractors? They are part of the team and want to participate. • Recommending making the event mandatory for the duration, with mission being the only acceptable reason to miss.[a]
Substantive nature of UNITE activities	Refers to comments that made recommendations about substantive aspects of UNITE activities and included such recommendations as using a facilitator or doing activities that allow for more interaction and direct communication between unit members. Some of these comments highlighted the importance of interaction between all unit members. Other comments highlighted a desire for units to have more leeway when planning events.	• I would like to see more unit involved things to connect with other [people] in our unit that are in different flights. • When we go to unit events everyone sticks to their co-workers. Maybe try and mix the groups up so people can get to know people in other sections. • Let Commanders be responsible for their people and plan their own events. • When a unit works 16hrs+ for a single shift, it's hard to get them motivated to participate in "forced fun" events, no matter how good the intentions of leadership may be. It may be better to allocate funds to a flight to organize their own events.

[a] However, some respondents thought that the "mandatory fun" feeling attached to UNITE events was a negative (see discussion of negative aspects of UNITE in this section), and at least one respondent explicitly said not to make UNITE events mandatory: "I feel as if these types of things shouldn't be mandatory, especially not if people don't want a donut."

Appendix C. Airman UNITE Survey Development and Administration

The UNITE Initiative is designed to provide unit commanders with opportunities to leverage FSS activities to increase unit cohesion. In turn, increased unit cohesion is expected to improve readiness and resilience among unit members. The Air Force asked RAND researchers to develop a survey, no longer than five minutes in length, that the Air Force could deploy to assess the initiative's effectiveness and contributions to unit cohesion. We developed two surveys: one to be administered immediately following UNITE participation (within two weeks of the event) and a second survey to be administered six weeks after the first survey.[35]

Our approach to designing a post-participation survey for the UNITE Initiative began with identifying which building blocks of readiness and resilience were most likely to be addressed by UNITE events. We also considered a variety of validated measures for the assessment of cohesion. In this appendix, we describe the process of survey development and implementation. The complete first and second surveys are provided at the end of this appendix.

Survey Development

We began by identifying the building blocks most commonly targeted by MWR programs, because we expected these to be similar to the building blocks most commonly targeted by FSS activities. We focused on outcomes and building blocks most likely to be affected by *unit* participation in FSS activities (e.g., we did not include such building blocks as family functioning). These are the short-term outcomes identified in the UNITE logic model in Chapter 2.

We then identified potential measures that could be used to define these outcomes, with adaptations made for the unit participation context. Because our goal was to design a very brief survey, we focused on single-item measures. Most of these outcomes were expected to be affected in the short term by UNITE participation (i.e., within the first two weeks after participation) and were measured in just the first survey. The exception was increased social interaction, which we anticipated might continue to improve following participation.

We developed an initial set of candidate questions, which were refined in consultation with staff from A1S and the Air Force Survey Office. A summary of the building blocks, outcomes, and survey items is presented in Table C.1. Note that certain outcomes map onto multiple building blocks in the model. Response options for each item were on a five-point scale that ranged from 1 (strongly disagree) to 5 (strongly agree).

[35] The surveys were an Air Force–approved data collection effort, Survey Control Number AF19-087A1S.

Table C.1. Building Blocks, Outcomes, and Survey Items

Readiness and Resilience Building Block(s)	Short-Term Outcome	Airman Post-Participation Survey Item
• Coping strategies and skills • Involvement in activities	• Positive use of leisure time	• Participating in this activity provided me with additional opportunities to interact or connect with members of my unit. (Survey 1) • Members of my unit are interacting and connecting more because of this activity. (Survey 2)
• Coping strategies and skills	• Opportunity to decompress	• This activity provided me with an opportunity to unwind (i.e., rest, relax, and/or have some fun).
• Peer group/unit values • Community/Air Force Values	• Promotion of Air Force institutional values	• Participating in this activity with unit members promoted or reinforced Air Force core values.
• Physical activity • Involvement in activities	• Increased physical activity	• Participating in this activity was physically demanding.
• Social network	• Increased social interaction	• Participating in this activity provided me with additional opportunities to interact or connect with members of my unit.

Cohesion Measures

Unlike observable measures that can be directly assessed and assigned a value—for example, how fast someone is, as measured by the time it takes them to run a mile—cohesion is a latent characteristic, something we measure indirectly. The literature on group cohesion provides multiple examples of survey-based measures used to assess group cohesion. Typically, these measures include subscales or factors to address the different components of cohesion, such as task and social cohesion. Task cohesion is the extent to which individual group members and the group as a whole are committed to the group task (i.e., whether group members help one another if they experience a problem while completing a task), whereas social cohesion is the extent to which individual group members and the group as a whole are attracted to the group social environment (i.e., whether members of the group socialize together; National Defense Research Institute, 2010).

The most commonly used scales are the GEQ (Carron, Widmeyer, and Brawley, 1985; Chang and Bordia, 2001; Gupta, Huang, and Niranjan, 2010; Widmeyer, Brawley, and Carron, 1985) and the Group Cohesion Scale–Revised (Treadwell et al., 2011). The GEQ is a general, rather than situation-specific, measure with four factors that aim to assess both task cohesion and social cohesion. The Group Cohesion Scale–Revised includes five factors: (1) interaction between group members, (2) sharing of information among group members, (3) decisionmaking among group members, (4) dependence among group members, and (5) group members sticking together.

Selected Cohesion Measures

We used the group integration subscales from the GEQ (Widmeyer, Brawley, and Carron, 1985), modified in ways similar to Chang and Bordia, 2001, and Gupta, Huang, and Niranjan, 2010, replacing the language of sports team, seasons, and game winnings to reflect Air Force unit, work hours, and tasks or mission. As Chang and Bordia, 2001, did, we removed the item "our team members have conflicting aspirations for the team's performance" from the group-integration task scale because it is not directly applicable to military units. Cohesion was measured at both time points, because we expected that changes in cohesion might take time to develop and, thus, observe following UNITE participation. The following is the final set of cohesion questions included in the UNITE survey; items with an asterisk are reverse scored:

- Group-integration social
  - Members of my unit would rather go out on their own than as a group.*
  - Members of my unit rarely socialize together.*
  - Members of my unit like to spend time together outside of work hours.
  - Members of my unit stick together outside of unit projects or tasks.
- Group-integration task
  - My unit is united in trying to accomplish its mission.
  - Members of my unit take responsibility for poor performance by our unit.
  - Members of my unit try to help if unit members have a problem with a task.
  - Members of my unit communicate freely about each other's responsibility.

Response options for each item were on a five-point scale that ranged from 1 (strongly disagree) to 5 (strongly agree). Group cohesion is calculated in a two-step process. The first step is to find the average response value of each subscale (i.e., group-integration social, group-integration task). Then, the two subscales are averaged for the individuals' overall group cohesion response value.

These cohesion measures had good internal reliability at both the first and second post-event survey, as measured by Cronbach's alpha. At the first post-event survey: social cohesion (alpha = 0.7836); task cohesion (alpha = 0.8588); overall unit cohesion (alpha = 0.8390). At the second post-event survey: social cohesion (alpha = 0.7839); task cohesion (alpha = 0.8645); overall unit cohesion (alpha = 0.8546).

Additional Items

We included two questions in addition to the items in the previous section. First, we included a question to determine whether an event took place during duty hours ("Did you participate in this activity during duty hours?"). Only participants who indicated that the event took place during off-duty hours received the question related to positive use of leisure time. We also included a single item to assess satisfaction with the activity ("I was satisfied with the activity my unit participated in."). Response options for the satisfaction item were on a five-point scale that ranged from 1 (strongly disagree) to 5 (strongly agree).

Implementing the UNITE Survey

Although the survey was developed in consultation with A1S and the Air Force Survey Office for the purposes of this study, it was ultimately designed to be fielded by A1S. This process allowed us to conduct an initial evaluation of UNITE and allowed for the survey to be used as an ongoing evaluation tool. The process of implementing the survey involved A1S, UNITE and AFSVC personnel, and AFPC.

Specifically, unit POCs at UNITE events were requested to collect DoD IDs from individuals who participated in the UNITE event for their unit. The list of DoD IDs was submitted to C3s, and C3s entered the DoD IDs into the C3 SharePoint database. AFSVC personnel then provided the Air Force Survey Office with the list of DoD IDs and relevant event details (i.e., name and date of event), which was uploaded into the survey system. AFPC used the provided DoD IDs to identify UNITE participant email addresses and issue personalized email invitations. AFPC administered the first survey every two weeks, sending an invitation to any individual who participated in a UNITE event within the previous two weeks. The second survey was sent approximately 45 days (or six weeks) after the initial survey invitation.

Prior to sharing data with A1S, AFPC linked survey responses with relevant demographic information about the individual from Air Force personnel records and then stripped the survey response of the DoD ID and email address. AFPC also provided A1S with deidentified data on the demographic characteristics of UNITE participants who were invited to take the post-participation surveys but who elected not to respond. The research team received deidentified data files from A1S via a secure file transmission.

Survey Invitations and Respondents

All active-duty airmen who participated in a UNITE event between July 27, 2019, and December 6, 2019, were eligible to complete a survey.[36] However, not all eligible airmen received a survey invitation, for the following reasons:

- **Collection of DoD IDs.** Not all participants were willing to provide their DoD ID to their unit POC and/or installation C3. We learned that, in some cases, these refusals were by individuals who did not feel comfortable providing the information. We also learned that, in other cases, unit or squadron commanders explicitly told their airmen not to provide the DoD ID to the C3s coordinating the event. Unfortunately, we were unable to identify

---

[36] Several UNITE events were held prior to the survey being approved by the relevant regulatory bodies. Therefore, individuals who participated in a UNITE event prior to July 27, 2019, received a survey invitation in early August, regardless of when the event occurred. Because the survey was designed to be administered in the two weeks following UNITE participation, we excluded responses from individuals who attended a UNITE event before July 27, 2019.

the number of UNITE participants who opted not to provide a DoD ID, because the data we had available did not enable us to identify the exact number of active-duty personnel who participated in each event (the number tracked in C3 data includes National Guard, Reserve, and civilians). This limits our ability to identify the number of active-duty personnel who did not get invited to the survey.

- **Accuracy of collected DoD IDs.** AFPC informed us that there was some data loss because of inaccurate DoD IDs. Specifically, many C3s were given handwritten lists of DoD IDs and entered those IDs into the C3 SharePoint database. Some IDs may have been incorrectly written down, or the handwriting may have been difficult to decipher. Thus, when AFPC received the list of DoD IDs, some of the IDs could not be matched to personnel records.

Invited Participants and Responding Participants

As with most surveys in civilian or military populations, not all individuals who received a personalized invitation to the survey chose to respond. Overall, we had a response rate of 14.2 percent for the initial follow-up survey and 8.5 percent for the second follow-up survey. We are able to determine differences in the survey respondents relative to the group that was invited. We extend this one step and also compare these groups to the active-duty Air Force (fiscal year 2018). The survey respondents in our analysis neither perfectly reflect the Air Force population nor reflect the group of individuals who participated in UNITE *and* were invited to participate in the survey. Table C.2 provides demographic characteristics across the fiscal year 2018 active-duty Air Force, airmen invited to take the first post-participation survey, and those who participated in the survey. Notably, the survey respondents were more likely to be female and of a higher rank. For example, active-duty enlisted males account for 64.4 percent of the Air Force and 63.5 percent of the group invited to participate in the survey. However, these individuals made up only 55.4 percent of the survey respondents and are thus underrepresented in our analysis. We also saw trends of airmen of lower enlisted ranks (e.g., E1–E4) responding at lower rates than airmen of higher enlisted ranks (e.g., E5–E6). Although we are unable to make our survey respondent population reflect the Air Force population or all active-duty personnel who participated in a UNITE event from the time of its inception, we include controls for the individual demographics to address possible sample bias.[37]

[37] Without complete or accurate DoD IDs for all individuals who participated in UNITE, we are unable to weight survey respondent data such that our analytic results, presented in Chapter 4 and Appendix D, would more closely reflect the perceptions of the current Air Force population or the active-duty Air Force population that participated in a UNITE event.

Table C.2. Demographic Characteristics of the Active-Duty Air Force Compared with the First UNITE Survey Invitees and Respondents

	Active-Duty Air Force (FY 2018)	UNITE Survey Invitee	UNITE Survey Respondent
Airman (E1–E3)	22.7%	17.7%	15.5%
Senior airman (E4)	16.0%	17.8%	9.3%
Noncommissioned officer (E5–E6)	20.2%	23.1%	29.3%
Senior noncommissioned officer (E7–E9)	4.9%	5.8%	8.7%
Company grade officer (O1–O3)	11.1%	11.4%	14.3%
Field grade officer (O4–O6)	4.0%	2.8%	5.4%
Female	20.2%	22.6%	26.1%
Enlisted	16.1%	18.8%	19.7%
Officer	4.1%	3.8%	6.4%
Male	79.8%	77.4%	73.9%
Enlisted	64.4%	63.5%	55.4%
Officer	15.3%	13.9%	18.5%
Number of persons	321,322	16,801	2,390

NOTE: FY = fiscal year. The active-duty Air Force demographic characteristics reflect the fiscal year 2018 Air Force (U.S. Department of Defense, 2019). The survey invitee and survey respondent populations reflect those individuals who participated in UNITE events that occurred between July 27, 2019, and December 10, 2019. Flag officers (O7 and above) are excluded.

Survey Instruments

In this section, we provide the first and second post-event participation survey instruments, including instructions and the required privacy act statements.

First Survey

Instructions

You received this survey because you recently participated in an Air Force-sponsored unit cohesion activity organized by your unit leadership. This survey is an opportunity to provide feedback about your experience and will only take 1-2 minutes to complete.

Completion of this survey is voluntary but highly encouraged to provide useful information to leadership.

While your survey responses will be kept confidential, summarized responses may be released to the public. Additionally, we cannot provide confidentiality to a participant regarding comments involving criminal activity/behavior, or statements that pose a threat to yourself or others. Do NOT discuss or comment on classified or operationally sensitive information at any point in this survey.

Privacy Act Statement

Authority: 10 U.S.C.; 8013, SECAF

Purpose: To better understand the effects of participating in unit cohesion activities.

Routine Uses: Feedback will be used to improve mission readiness.
Disclosure: Providing information in this survey is voluntary. Individual responses will not be shared with others unless required by law. The survey results are reported only in a manner that does not identify an individual.

UNITE Screening Question

According to our records, you recently participated in the following Air Force-sponsored unit cohesion activity organized by your unit leadership:

Activity Description: [Filled from data submitted by C3]

Date: [Filled from data submitted by C3]

Is this correct?

- ☐ Yes
- ☐ No

[IF NO] [*End Survey]*

[IF YES] When answering survey questions, "activity" refers to the following:

Activity Description: [Filled from data submitted by C3]

Date: [Filled from data submitted by C3]

Did you participate in this activity during duty hours?

- ☐ Yes
- ☐ No

Please indicate how much you agree or disagree with each of the following statements.

	Strongly disagree	Disagree	Neither agree nor disagree	Agree	Strongly agree	I do not wish to answer
I was satisfied with the activity my unit participated in.	1	2	3	4	5	6
Participating in this activity was physically demanding.	1	2	3	4	5	6
[IF PARTICIPANT INDICATED THEY DID NOT PARTICIPATE DURING DUTY HOURS] Participating in this activity was an enjoyable use of my free time.	1	2	3	4	5	6
This activity provided me with an opportunity to unwind (i.e., rest, relax, and/or have some fun).	1	2	3	4	5	6
Participating in this activity provided me with additional opportunities to interact or connect with members of my unit.	1	2	3	4	5	6

Participating in this activity with unit members promoted or reinforced Air Force core values.	1	2	3	4	5	6
Members of my unit would rather go out on their own than as a group.	1	2	3	4	5	6
Members of my unit rarely socialize together.	1	2	3	4	5	6
Members of my unit like to spend time together outside of work hours.	1	2	3	4	5	6
Members of my unit stick together outside of unit projects or tasks.	1	2	3	4	5	6
My unit is united in trying to accomplish its mission.	1	2	3	4	5	6
Members of my unit take responsibility for poor performance by our unit.	1	2	3	4	5	6
Members of my unit try to help if unit members have a problem with a task.	1	2	3	4	5	6
Members of my unit communicate freely about each other's responsibility.	1	2	3	4	5	6

Do you have any other general comments you would like to provide?
We cannot provide confidentiality to a participant regarding comments involving criminal activity/behavior, or statements that pose a threat to yourself or others. Do NOT discuss or comment on classified or operationally sensitive information.

[FREE RESPONSE]

[*End First Survey]*

Second Survey

Instructions/Start Screen
You received this survey because you participated in an Air Force-sponsored unit cohesion activity organized by your unit leadership some time ago. This survey is an opportunity to provide feedback about your experience and will only take 1-2 minutes to complete. Note that this is a different survey than you were previously contacted about, though some of the questions may appear similar.

Completion of this survey is voluntary but highly encouraged to provide useful information to leadership.

While your survey responses will be kept confidential, summarized responses may be released to the public. Additionally, we cannot provide confidentiality to a participant regarding comments involving criminal activity/behavior, or statements that pose a threat to yourself or others. Do NOT discuss or comment on classified or operationally sensitive information at any point in this survey.

Privacy Act Statement
Authority: 10 U.S.C.; 8013, SECAF

Purpose: To better understand the effects of participating in unit cohesion activities.
Routine Uses: Feedback will be used to improve mission readiness.
Disclosure: Providing information in this survey is voluntary. Individual responses will not be shared with others unless required by law. The survey results are reported only in a manner that does not identify an individual.

Screening

According to our records, you recently participated in the following Air Force-sponsored unit cohesion activity organized by your unit leadership:

Activity Description: [Filled from data submitted by C3]

Date: [Filled from data submitted by C3]

Is this correct?

☐ Yes
☐ No

[IF YES] When answering survey questions, "activity" refers to the following:

Activity Description: [Filled from data submitted by C3]

Date: [Filled from data submitted by C3]

Did you participate in this activity during duty hours?

☐ Yes
☐ No

Please indicate how much you agree or disagree with each of the following statements.

	Strongly disagree	Disagree	Neither agree nor disagree	Agree	Strongly agree	I do not wish to answer
Members of my unit are interacting and connecting more because of this activity.	1	2	3	4	5	6
Members of my unit would rather go out on their own than as a group.	1	2	3	4	5	6
Members of my unit rarely socialize together.	1	2	3	4	5	6
Members of my unit like to spend time together outside of work hours.	1	2	3	4	5	6
Members of my unit stick together outside of unit projects or tasks.	1	2	3	4	5	6
My unit is united in trying to accomplish its mission.	1	2	3	4	5	6
Members of my unit take responsibility for poor performance by our unit.	1	2	3	4	5	6

Members of my unit try to help if unit members have a problem with a task.	1	2	3	4	5	6
Members of my unit communicate freely about each other's responsibility.	1	2	3	4	5	6

Do you have any other general comments you would like to provide?
We cannot provide confidentiality to a participant regarding comments involving criminal activity/behavior, or statements that pose a threat to yourself or others. Do NOT discuss or comment on classified or operationally sensitive information.

[FREE RESPONSE]

Thank you for taking the time to complete this survey.

[End Second Survey]

Appendix D. Quantitative Evaluation Methods: Outcome Evaluation

In this appendix, we review the analytic approach used to estimate the association between UNITE and unit cohesion. We review the main research questions the overall study was designed to address, the data used for the analysis, and the statistical approaches leveraged. Because of the nature of the available data, all planned analyses are correlational, and causality cannot be inferred from the results. That is, we are not able to determine whether participation in a UNITE event caused a change in unit cohesion.

Research Questions

We explored three research questions regarding the associations between UNITE and unit cohesion:

1. How are event characteristics directly associated with the readiness and resilience building blocks?
2. How are readiness and resilience building blocks directly and indirectly associated with unit, social, and task cohesion at both the first and second post-event surveys?
3. How are event characteristics associated with unit, social, and task cohesion, both directly and indirectly, through the readiness and resilience building blocks?

Data

We leveraged multiple sources of data for our analyses, including the following:

- Air Force–provided data on UNITE participant characteristics (gender, age, and rank).
    - We grouped ranks into the following: airman (E1–E3), senior airman (E4), noncommissioned officer (E5–E6), senior noncommissioned officer (E7–E9), company grade officer (O1–O3), and field grade officer (O4–O6).
- Air Force–provided data on installation characteristics (remote/isolated status and size of active-duty population on an installation).
- Air Force–provided responses on the post-event surveys (see Appendix C). The first post-UNITE participation survey was administered to UNITE participants two weeks after participating in a UNITE event and included such questions as whether the event occurred during duty hours and questions that assessed the building blocks of readiness and resilience, satisfaction with the specific UNITE event they participated in, and the GEQ measure of task and social cohesion. The second post-event survey was administered approximately eight weeks after participating in a UNITE event and included the GEQ cohesion scale.

- Air Force–provided C3 SharePoint database. C3s provide data on each UNITE event implemented, including date, location (on or off base), whether the event was provided through the installation's FSS, and the types of funding used for the event.

Statistical Analyses

We employed path analysis to answer our research questions. With path analyses, we model the associations between event characteristics, building blocks, and cohesion outcomes from the first and second post-event surveys. As discussed in the main body of the report, Figure D.1 illustrates a simplified path diagram.

Figure D.1. Simplified Path Analysis Model

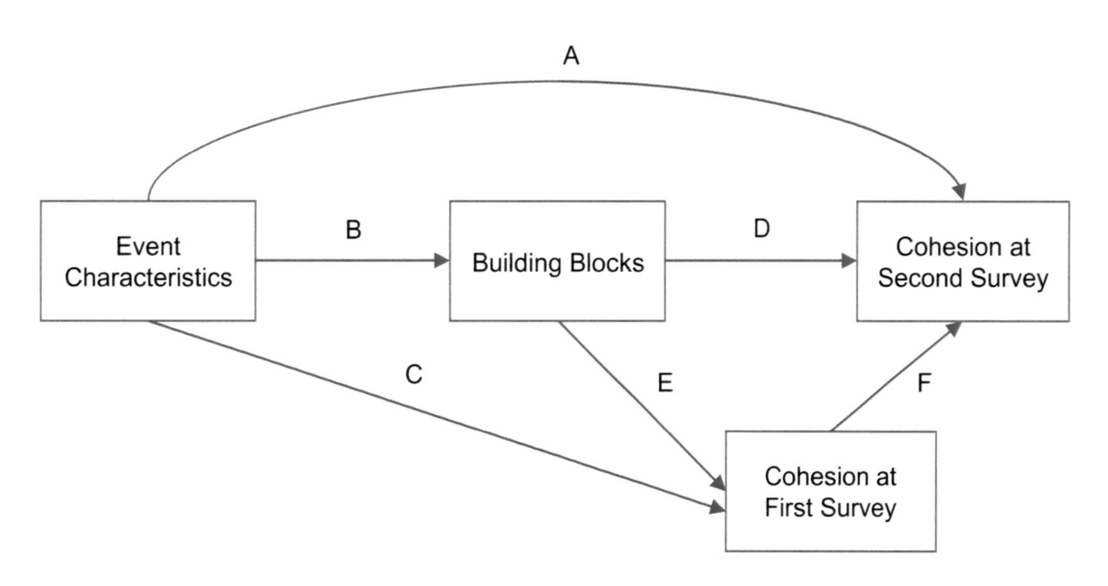

Path models allow for the estimation of direct associations (e.g., the direct association of event characteristics on cohesion from the second post-event survey, represented by path A) and indirect associations (e.g., the indirect association of event characteristics on cohesion from the second post-event survey through building blocks, represented by paths B and D). All possible paths are estimated simultaneously with multivariate regressions.

Event characteristics included in the model were indicators for whether the event used MOA funds and whether the event was provided by the installation FSS. Our final model did not include indicators for an event occurring during duty hours or off base because analyses revealed no significant direct associations with those event characteristics and any building block or cohesion outcome. All building blocks, except physical activity, were included in the model. Physical activity was omitted because there were no direct associations between physical activity and any event characteristic or cohesion outcome.

We allowed for all possible direct associations between event characteristics and building blocks, event characteristics and cohesion outcomes, and building blocks and cohesion outcomes. All models also control for gender, age, rank, being assigned to a remote or isolated installation, and installation size. Standard errors are clustered by installation.

In this appendix, we present results for three outcomes: task cohesion, social cohesion, and overall unit cohesion (the average of task and social cohesion). Further, we present path diagrams that include only paths with significant associations. The tables in this appendix include all direct associations, indirect associations, and total associations, regardless of significance.

Research Question 1

Our first question asks how event characteristics are directly associated with the readiness and resilience building blocks. In the path diagram illustrated in Figure D.1, the answer to this question is given by path B. Generally, the regression can be written as

$$BB_{ibt} = \beta_0 + \beta_B EC_{ibt}\beta_1 + \boldsymbol{X_{ibt}\beta_2} + e1_{ibt},$$

where BB_{ibt} represents one of the three building block measures (decompress, social interaction, or peer/Air Force values) for individual *i* on installation *b* at time *t*. EC_{ibt} is one of the event characteristics (i.e., MOA-funded or FSS provided), $\boldsymbol{X_{ibt}}$ is a vector of aforementioned airmen and installation characteristics, and $e1_{ibt}$ is an idiosyncratic individual-level error term. The coefficient of interest is β_B, which estimates the direct association of each building block and each event characteristic as represented by path B. In total, six equations were estimated, one for each event characteristic and building block combination.

Research Question 2

Our second research question asks how readiness and resilience building blocks are directly and indirectly associated with unit, social, and task cohesion at both the first and second post-event surveys. In Figure D.1, the direct associations between building blocks and first post-event surveys are given by path E, which can be modeled as follows:

$$C1_{ibt} = \beta_E BB_{ibt} + \boldsymbol{X_{ibt}\beta_2} + e2_{ibt},$$

where $C1_{ibt}$ represents the overall social or task cohesion measured from the first post-event survey, BB_{ibt} represents one of the three building blocks, and $\boldsymbol{X_{ibt}}$ represents the same vector of control variables. Three equations are estimated, one for each building block. It follows that the direct association with cohesion at the second post-event survey can be modeled with the following analogous set of equations:

$$C2_{ibt} = \beta_D BB_{ibt} + \boldsymbol{X_{ibt}\beta_2} + e3_{ibt},$$

where $C2_{ibt}$ represents the cohesion measure on the second post-event survey.

To estimate the total association of building blocks and cohesion at the second post-event survey, we combined the direct association (path D) with the indirect association through cohesion from the first post-event survey (paths E and F). The total association can be modeled as

$$C2_{ibt} = (\beta_D + \beta_E\beta_F)BB_{ibt} + \boldsymbol{X_{ibt}\beta_2} + e4_{ibt}.$$

Finally, to understand whether the association with cohesion persists from the first to the second post-event survey, we compared β_E (i.e., the direct association with cohesion from the first post-event survey) and $\beta_D + \beta_E\beta_F$ (i.e., the total association with cohesion from the second post-event survey).

Research Question 3

Our third and last research question asks how event characteristics are associated with unit, social, and task cohesion (both directly and indirectly) through the readiness and resilience building blocks. It follows from the models used in research questions 1 and 2 that the following equations can be used to answer part of the question.

Cohesion at the First Post-Event Survey

$$\text{Direct association: } C1_{ibt} = \beta_C EC_{ibt} + \boldsymbol{X_{ibt}\beta_2} + e5_{ibt}$$
$$\text{Indirect association: } C1_{ibt} = \beta_B\beta_E EC_{ibt} + \boldsymbol{X_{ibt}\beta_2} + e6_{ibt}$$
$$\text{Total association: } C1_{ibt} = (\beta_C + \beta_B\beta_E)EC_{ibt} + \boldsymbol{X_{ibt}\beta_2} + e7_{ibt}$$

Cohesion at the Second Post-Event Survey

$$\text{Direct association: } C2_{ibt} = \beta_A EC_{ibt} + \boldsymbol{X_{ibt}\beta_2} + e8_{ibt}$$
$$\text{Indirect association: } C2_{ibt} = (\beta_B\beta_D + \beta_C\beta_F + \beta_B\beta_E\beta_F)EC_{ibt} + \boldsymbol{X_{ibt}\beta_2} + e9_{ibt}$$
$$\text{Total association: } C2_{ibt} = (\beta_A + \beta_B\beta_D + \beta_C\beta_F + \beta_B\beta_E\beta_F)EC_{ibt} + \boldsymbol{X_{ibt}\beta_2} + e10_{ibt}$$

When we present our results, path diagrams show all significant direct associations, and tables include all direct, indirect, and total associations, regardless of significance level.

Full Results

We discuss key aspects of our analytic results in Chapter Four, and we provide the full set of covariates and results in this section. We begin with basic descriptive characteristics of our samples. The full sample of airmen who completed the first post-event survey and the subsample who completed both the first and second post-event surveys look largely similar, as shown in Table D.1.

Table D.1. Descriptive Characteristics of Airmen in the Analytic Samples

	First Post-Event Survey (2,216 Respondents)		First and Second Post-Event Surveys (624 Respondents)	
	Mean[a]	SD	Mean[a]	SD
Panel A: Airmen and Military Demographics				
Male	73.8%	--	76.1%	--
Age	31.392	7.324	32.189	7.669
Rank (pay grade)				
Airman (E1–E3)	15.5%	--	15.5%	--
Senior airman (E4)	8.9%	--	7.1%	--
Noncommissioned officer (E5–E6)	31.4%	--	29.0%	--
Senior noncommissioned officer (E7–E9)	19.1%	--	20.2%	--
Company grade officer (O1–O3)	14.4%	--	13.3%	--
Field grade officer (O4–O6)	10.7%	--	14.9%	--
Panel B: Installation Characteristics				
Assigned to remote/isolated	23.0%	--	22.3%	--
Installation size				
Small (up to 1,000 active-duty airmen)	7.0%	--	6.7%	--
Medium (1,001–5,000 active-duty airmen)	65.6%	--	69.7%	--
Large (5,001–6,000 active-duty airmen)	10.4%	--	12.3%	--
Mega-large (6,001–30,000 active-duty airmen)	14.0%	--	11.2%	--
Panel C: Event Characteristics				
During duty hours	85.8%	--	85.3%	--
Off base	46.9%	--	47.4%	--
MOA-funded	69.4%	--	68.1%	--
FSS-provided	51.5%	--	51.9%	--
Satisfaction	4.526	0.827	4.577	--
Commander reasons for UNITE event				
To provide opportunity for fun or relaxation	91.2%	--	92.6%	--
To promote interaction between unit members	94.6%	--	94.5%	--
To increase morale, camaraderie, or esprit de corps	97.0%	--	97.6%	--
To improve physical fitness	37.8%	--	36.6%	--
To work on a team-building exercise	59.4%	--	57.8%	--
To develop a new skill or competency	31.9%	--	31.1%	--
To reinforce peer, unit/squadron, or Air Force values	59.6%	--	57.9%	--

	First Post-Event Survey (2,216 Respondents)		First and Second Post-Event Surveys (624 Respondents)	
	Mean[a]	SD	Mean[a]	SD
Panel D: Building Blocks				
Physical activity	2.475	1.145	2.497	1.16
Positive use of leisure time (off-duty events only)	4.341	1.049	4.48	0.955
Coping skills and strategy: involvement in activities (decompress)	4.371	0.933	4.413	0.893
Social network (social interaction), Survey 1	4.484	0.812	4.527	0.787
Social network (social interaction), Survey 2	--	--	3.923	0.968
Peer/Air Force values	4.066	0.983	4.088	0.96
Panel E: Cohesion				
Social cohesion, Survey 1	3.236	0.744	3.281	0.738
Task cohesion, Survey 1	4.002	0.76	4.055	0.748
Overall cohesion, Survey 1	3.619	0.638	3.668	0.638
Social cohesion, Survey 2			3.338	0.762
Task cohesion, Survey 2			4.123	0.751
Overall cohesion, Survey 2			3.731	0.657

NOTE: SD = standard deviation.
[a] Categorical variables are presented as percentages without accompanying standard deviations.

Main Path Diagrams on All Cohesion Outcomes

The path diagrams in the main body of the report present the results only for the overall cohesion outcome. Here we present this result again, along with analogous diagrams in which task and social cohesion were the outcomes of interest. Figure D.2 is a reproduction of Figure 4.4 and shows that there were no significant direct associations between event characteristics and cohesion outcomes or direct associations between building blocks and cohesion from the second post-event surveys. However, there were direct associations between event characteristics and building blocks and building blocks and cohesion at the first post-event survey.

Figure D.2. Path Diagram of Event Characteristics, Building Blocks, and Overall Cohesion

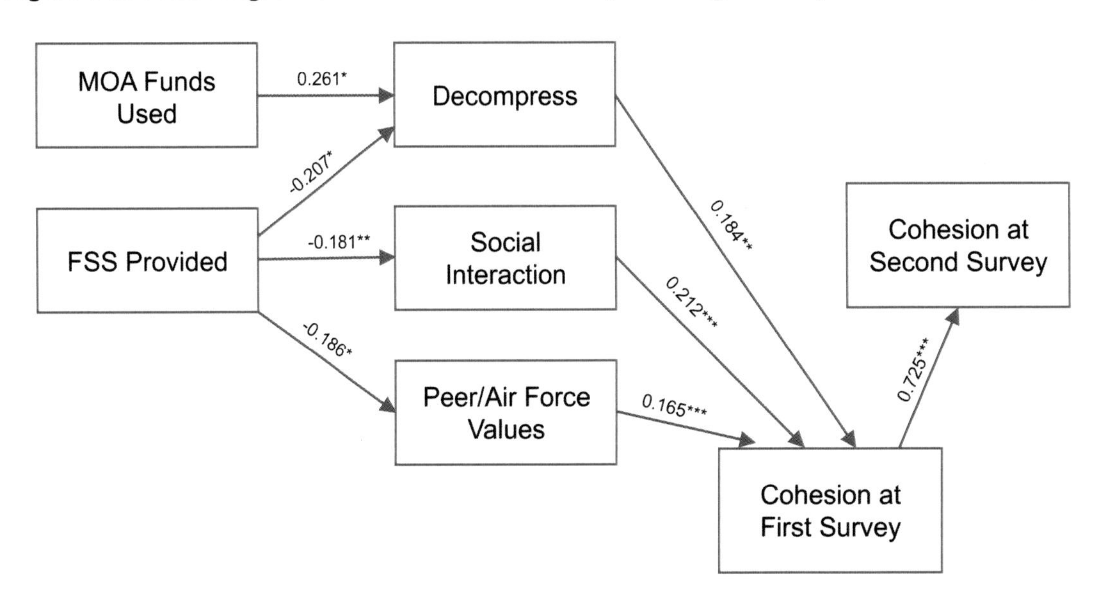

NOTE: $N = 624$. Only paths with significant associations are illustrated. The full model allows for all possible paths between event characteristics, building blocks, and cohesion outcomes, as well as paths between building blocks and cohesion from the second post-event survey. Standard errors (not shown) are clustered at the installation level. Models control for gender, age, rank, remote or isolated installation, and installation size. * indicates $p < 0.05$; ** indicates $p < 0.01$; and *** indicates $p < 0.001$.

Figure D.3 shows the analogous path diagram with task cohesion as the outcome. Direct associations between event characteristics and building blocks were identical. As with overall cohesion, there were no direct associations between event characteristics and cohesion outcomes or direct associations between building blocks and cohesion from the second post-event survey. Further, direct associations between building blocks and cohesion outcomes were of similar magnitude and significance.

Figure D.3. Path Diagram of Event Characteristics, Building Blocks, and Task Cohesion

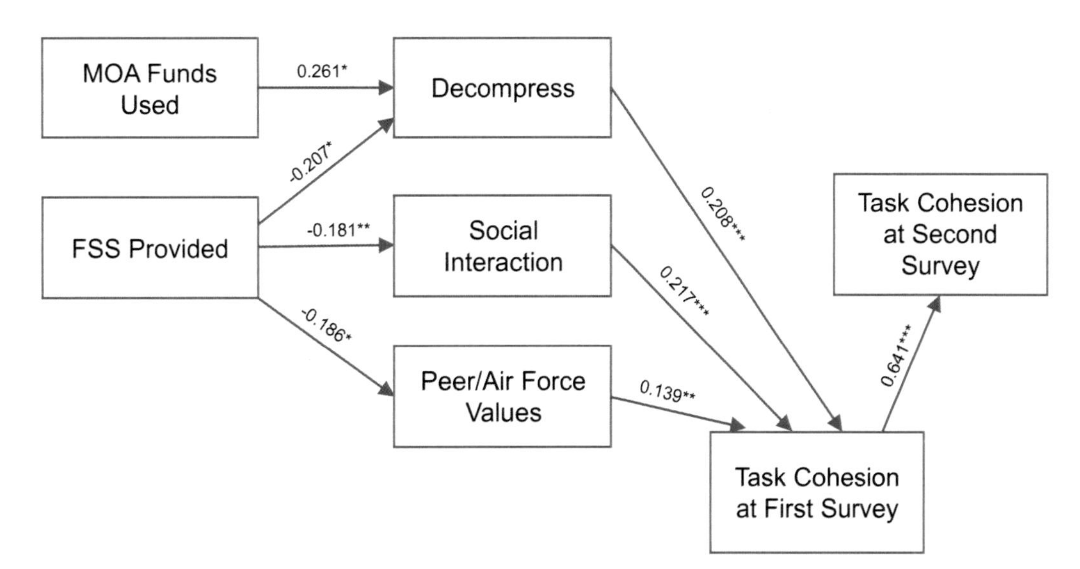

NOTE: $N = 624$. Only paths with significant associations are illustrated. The full model allows for all possible paths between event characteristics, building blocks, and cohesion outcomes, as well as paths between building blocks and cohesion from the second post-event survey. Standard errors (not shown) are clustered at the installation level. Models control for gender, age, rank, remote or isolated installation, and installation size. * indicates $p < 0.05$; ** indicates $p < 0.01$; and *** indicates $p < 0.001$.

Finally, Figure D.4 shows the analogous path diagram with social cohesion as the outcome. Again, the direct associations between event characteristics and building blocks were identical to previous models. However, there was a slight difference in the direct associations between building blocks and the social cohesion outcome. Here, the decompression building block no longer had a direct association with cohesion from the first post-event survey and instead had a direct association with cohesion from the second post-event survey.

Figure D.4. Path Diagram of Event Characteristics, Building Blocks, and Social Cohesion

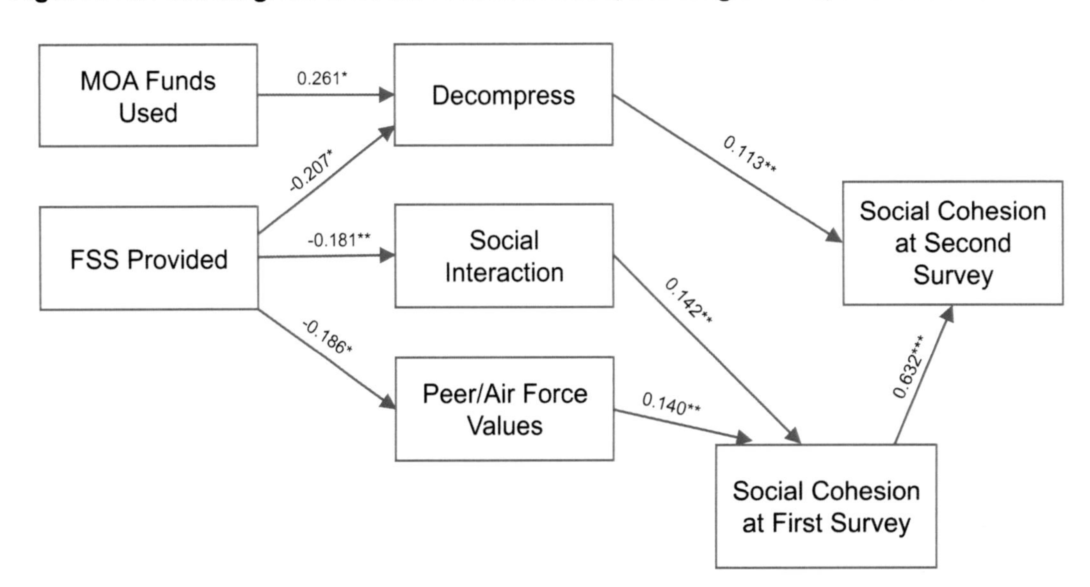

NOTE: N = 624. Only paths with significant associations are illustrated. The full model allows for all possible paths between event characteristics, building blocks, and cohesion outcomes, as well as paths between building blocks and cohesion from the second post-event survey. Standard errors (not shown) are clustered at the installation level. Models control for gender, age, rank, remote or isolated installation, and installation size. * indicates $p < 0.05$; ** indicates $p < 0.01$; and *** indicates $p < 0.001$.

Tables of All Direct, Indirect, and Total Associations from Main Models

As stated earlier, the figures in the previous section and those in the main body of the report contain only paths that had significant associations. Next, we present estimates for all paths, regardless of significance, as well as estimates for the control variables. We also present results for the three types of cohesion outcomes: overall unit cohesion, task cohesion, and social cohesion. For each measure of cohesion, we present three types of estimates: (1) direct associations between event characteristics, building blocks, and cohesion outcomes, (2) indirect associations among the three groups of variables, and (3) total associations among the three groups of variables.

Table D.2 presents direct associations of event characteristics, building blocks, and overall unit cohesion. Significant path associations from this table are found in Figures 4.4 and D.3. Table D.3 presents indirect and total associations of event characteristics, building blocks and overall cohesion. Some estimates in this table are found in Tables 4.5 and 4.6. Table D.4 presents direct associations between event characteristics, building blocks, and task cohesion, and Table D.5 presents the analogous indirect and total associations. Finally, Table D.6 presents the direct associations between event characteristics, building blocks, and social cohesion, and Table D.7 presents the analogous indirect and total associations.

Table D.2. Direct Associations of Event Characteristics, Building Blocks, and Overall Cohesion

	Cohesion First Post-Event Survey Direct Associations	Cohesion Second Post-Event Survey Direct Associations	Decompress Direct Associations	Social Interaction Direct Associations	Peer/Air Force Values Direct Associations
Decompress	0.184**	0.052			
	(0.058)	(0.048)			
Social interaction	0.212***	−0.083			
	(0.051)	(0.053)			
Peer/Air Force values	0.165***	0.029			
	(0.045)	(0.035)			
MOA-funded	−0.057	0.089	0.261*	0.117	0.071
	(0.100)	(0.049)	(0.107)	(0.098)	(0.070)
FSS-provided	0.056	−0.010	−0.207*	−0.181**	−0.186*
	(0.059)	(0.066)	(0.089)	(0.061)	(0.079)
Male	0.258**	0.034	0.024	−0.037	0.204*
	(0.097)	(0.056)	(0.082)	(0.084)	(0.092)
Age	−0.002	0.003	0.008	−0.001	0.008
	(0.008)	(0.005)	(0.008)	(0.009)	(0.008)
Airman (E1–E3)	0.503***	−0.083	0.279*	−0.132	0.106
	(0.130)	(0.095)	(0.128)	(0.147)	(0.142)
Senior airman (E4)	0.083	−0.050	0.302	−0.003	0.206
	(0.198)	(0.106)	(0.186)	(0.181)	(0.170)
Senior noncommissioned officer (E7–E9)	0.224*	−0.059	0.220*	0.145	0.223*
	(0.099)	(0.085)	(0.111)	(0.120)	(0.102)
Company grade officer (O1–O3)	0.310*	0.085	0.270*	0.197	0.248
	(0.134)	(0.076)	(0.130)	(0.119)	(0.131)
Field grade officer (O4–O6)	0.519***	0.148	0.250*	0.360**	0.276*
	(0.115)	(0.109)	(0.115)	(0.120)	(0.137)
Remote/isolated installation	0.152	0.082	−0.108	0.018	−0.098
	(0.110)	(0.061)	(0.146)	(0.097)	(0.087)
Installation of medium size	0.159	−0.057	0.121	0.096	−0.034
	(0.128)	(0.056)	(0.133)	(0.127)	(0.118)
Installation of large size	0.286*	−0.048	0.115	0.137	0.060
	(0.143)	(0.080)	(0.171)	(0.182)	(0.146)
Installation of mega-large size	−0.083	−0.117	0.119	0.023	0.101
	(0.186)	(0.161)	(0.185)	(0.169)	(0.129)

	Cohesion First Post-Event Survey Direct Associations	Cohesion Second Post-Event Survey Direct Associations	Decompress Direct Associations	Social Interaction Direct Associations	Peer/Air Force Values Direct Associations
Cohesion, first post-event survey		0.725***			
		(0.042)			
Constant	−0.490	−0.195	−0.581*	−0.035	−0.455
	(0.317)	(0.163)	(0.284)	(0.291)	(0.291)
Observations	624	624	624	624	624

NOTE: Standard errors are clustered at the installation level. All building blocks and cohesion outcomes are standardized. Reference category includes noncommissioned officer (E7–E9), non-remote/isolated installation, and small installation size. * indicates $p < 0.05$; ** indicates $p < 0.01$; and *** indicates $p < 0.001$.

Table D.3. Indirect and Total Associations of Event Characteristics, Building Blocks, and Overall Unit Cohesion

	Overall Cohesion First Post-Event Survey Indirect Associations	Overall Cohesion Second Post-Event Survey Indirect Associations	Overall Cohesion First Post-Event Survey Total Associations	Overall Cohesion Second Post-Event Survey Total Associations
Decompress		0.133**	0.184**	0.185***
		(0.043)	(0.058)	(0.053)
Social interaction		0.154***	0.212***	0.071
		(0.038)	(0.051)	(0.062)
Peer/Air Force values		0.119***	0.165***	0.148**
		(0.032)	(0.045)	(0.051)
MOA-funded	0.084	0.025	0.027	0.115
	(0.045)	(0.082)	(0.113)	(0.099)
FSS-provided	−0.107**	−0.038	−0.051	−0.048
	(0.035)	(0.056)	(0.076)	(0.086)
Male	0.030	0.219**	0.288**	0.254**
	(0.039)	(0.076)	(0.107)	(0.094)
Age	0.002	0.001	0.001	0.004
	(0.004)	(0.006)	(0.0088)	(0.008)
Airman (E1–E3)	0.041	0.422***	0.544***	0.339*
	(0.073)	(0.104)	(0.151)	(0.155)
Senior airman (E4)	0.089	0.146	0.172	0.097
	(0.093)	(0.152)	(0.214)	(0.193)
Senior noncommissioned officer (E7–E9)	0.108*	0.246**	0.332**	0.187
	(0.052)	(0.086)	(0.120)	(0.122)

	Overall Cohesion First Post-Event Survey Indirect Associations	Overall Cohesion Second Post-Event Survey Indirect Associations	Overall Cohesion First Post-Event Survey Total Associations	Overall Cohesion Second Post-Event Survey Total Associations
Company grade officer (O1–O3)	0.132*	0.325**	0.442**	0.410**
	(0.061)	(0.104)	(0.146)	(0.135)
Field grade officer (O4–O6)	0.168**	0.488***	0.686***	0.637
	(0.064)	(0.106)	(0.139)	(0.138)
Remote/isolated installation	−0.032	0.077	0.120	0.159
	(0.053)	(0.092)	(0.121)	(0.105)
Installation of medium size	0.037	0.140	0.196	0.083
	(0.057)	(0.122)	(0.163)	(0.139)
Installation of large size	0.060	0.247*	0.346*	0.200
	(0.083)	(0.122)	(0.163)	(0.148)
Installation of mega-large size	0.043	−0.021	−0.039	−0.138
	(0.081)	(0.165)	(0.225)	(0.157)
Cohesion, first post-event survey				0.725***
				(0.042)
Observations	624	624	624	624

NOTE: Standard errors are clustered at the installation level. All building blocks and cohesion outcomes are standardized. Reference category includes noncommissioned officer (E7–E9), non-remote/isolated installation, and small installation size. * indicates $p < 0.05$; ** indicates $p < 0.01$; and *** indicates $p < 0.001$.

Table D.4. Direct Associations of Event Characteristics, Building Blocks, and Task Cohesion

	Task Cohesion First Post-UNITE Survey Direct Associations	Task Cohesion Second Post-UNITE Survey Direct Associations	Decompress Direct Associations	Social Interaction Direct Associations	Peer/Air Force Values Direct Associations
Decompress	0.208***	0.010			
	(0.057)	(0.052)			
Social interaction	0.217***	−0.021			
	(0.063)	(0.055)			
Peer/Air Force values	0.139**	0.023			
	(0.043)	(0.036)			
MOA-funded	−0.091	0.087	0.261*	0.117	0.071
	(0.101)	(0.058)	(0.107)	(0.098)	(0.070)
FSS-provided	0.011	−0.032	−0.207*	−0.181**	−0.186*
	(0.068)	(0.070)	(0.089)	(0.061)	(0.079)

	Task Cohesion First Post-UNITE Survey Direct Associations	Task Cohesion Second Post-UNITE Survey Direct Associations	Decompress Direct Associations	Social Interaction Direct Associations	Peer/Air Force Values Direct Associations
Male	0.215*	0.118	0.024	−0.037	0.204*
	(0.096)	(0.067)	(0.082)	(0.084)	(0.092)
Age	0.004	0.004	0.008	−0.001	0.008
	(0.008)	(0.007)	(0.008)	(0.009)	(0.008)
Airman (E1–E3)	0.356**	−0.001	0.279*	−0.132	0.106
	(0.125)	(0.116)	(0.128)	(0.147)	(0.142)
Senior airman (E4)	0.052	−0.018	0.302	−0.003	0.206
	(0.183)	(0.133)	(0.186)	(0.181)	(0.170)
Senior noncommissioned officer (E7–E9)	0.105	0.043	0.220*	0.145	0.223*
	(0.104)	(0.093)	(0.111)	(0.120)	(0.102)
Company grade officer (O1–O3)	0.168	0.100	0.270*	0.197	0.248
	(0.129)	(0.096)	(0.130)	(0.119)	(0.131)
Field grade officer (O4–O6)	0.318**	0.190	0.250*	0.360**	0.276*
	(0.120)	(0.110)	(0.115)	(0.120)	(0.137)
Remote/isolated installation	0.031	−0.023	−0.108	0.018	−0.098
	(0.094)	(0.069)	(0.146)	(0.097)	(0.087)
Installation of medium size	0.104	−0.129	0.121	0.096	−0.034
	(0.136)	(0.085)	(0.133)	(0.127)	(0.118)
Installation of large size	0.117	−0.129	0.115	0.137	0.060
	(0.173)	(0.121)	(0.171)	(0.182)	(0.146)
Installation of mega-large size	−0.103	−0.227	0.119	0.023	0.101
	(0.184)	(0.156)	(0.185)	(0.169)	(0.129)
Cohesion, first post-event survey		0.641***			
		(0.043)			
Constant	−0.426	−0.214	−0.581*	−0.035	−0.455
	(0.332)	(0.216)	(0.284)	(0.291)	(0.291)
Observations	624	624	624	624	624

NOTE: Standard errors are clustered at the installation level. All building blocks and cohesion outcomes are standardized. Reference category includes noncommissioned officer (E7–E9), non-remote/isolated installation, and small installation size. * indicates $p < 0.05$; ** indicates $p < 0.01$; and *** indicates $p < 0.001$.

Table D.5. Indirect and Total Associations of Event Characteristics, Building Blocks, and Task Cohesion

	Task Cohesion First Post-UNITE Survey Indirect Associations	Task Cohesion Second Post-UNITE Survey Indirect Associations	Task Cohesion First Post-UNITE Survey Total Associations	Task Cohesion Second Post-UNITE Survey Total Associations
Decompress		0.133	0.208***	0.143*
		(0.037)	(0.057)	(0.057)
Social interaction		0.139	0.217**	0.118*
		(0.041)	(0.063)	(0.056)
Peer/Air Force values		0.089	0.139**	0.112*
		(0.029)	(0.043)	(0.053)
MOA-funded	0.089	0.001	−0.002	0.088
	(0.047)	(0.074)	(0.115)	(0.080)
FSS-provided	−0.108**	−0.065	−0.098	−0.097
	(0.035)	(0.052)	(0.079)	(0.085)
Male	0.025	0.159	0.240*	0.277**
	(0.039)	(0.068)	(0.103)	(0.100)
Age	0.002	0.005	0.007	0.008
	(0.004)	(0.006)	(0.009)	(0.009)
Airman (E1–E3)	0.044	0.264	0.400**	0.263
	(0.074)	(0.096)	(0.143)	(0.144)
Senior airman (E4)	0.091	0.099	0.143	0.081
	(0.096)	(0.142)	(0.218)	(0.188)
Senior noncommissioned officer (E7–E9)	0.108	0.141	0.213	−0.184
	(0.057)	(0.081)	(0.125)	(0.134)
Company grade officer (O1–O3)	0.133*	0.197	0.302*	0.298*
	(0.062)	(0.097)	(0.144)	(0.138)
Field grade officer (O4–O6)	0.169**	0.313	0.486**	0.503***
	(0.063)	(0.094)	(0.143)	(0.137)
Remote/isolated installation	−0.032	−0.004	−0.001	−0.027
	(0.055)	(0.072)	(0.112)	(0.083)
Installation of medium size	0.041	0.091	0.145	−0.038
	(0.058)	(0.100)	(0.152)	(0.148)
Installation of large size	0.062	0.114	0.179	−0.015
	(0.084)	(0.110)	(0.166)	(0.184)
Installation of mega-large size	0.044	−0.035	−0.059	−0.262
	(0.084)	(0.143)	(0.221)	(0.172)

	Task Cohesion First Post-UNITE Survey Indirect Associations	Task Cohesion Second Post-UNITE Survey Indirect Associations	Task Cohesion First Post-UNITE Survey Total Associations	Task Cohesion Second Post-UNITE Survey Total Associations
Cohesion, first post-event survey				0.641***
				(0.043)
Observations	624	624	624	624

NOTE: Standard errors are clustered at the installation level. All building blocks and cohesion outcomes are standardized. Reference category includes noncommissioned officer (E7–E9), non-remote/isolated installation, and small installation size. * indicates $p < 0.05$; ** indicates $p < 0.01$; and *** indicates $p < 0.001$.

Table D.6. Direct Associations of Event Characteristics, Building Blocks, and Social Cohesion

	Social Cohesion First Post-UNITE Survey Direct Associations	Social Cohesion Second Post-UNITE Survey Direct Associations	Decompress Direct Associations	Social Interaction Direct Associations	Peer/Air Force Values Direct Associations
Decompress	0.103	0.113**			
	(0.074)	(0.043)			
Social interaction	0.142**	−0.084			
	(0.054)	(0.050)			
Peer/Air Force values	0.140**	0.057			
	(0.052)	(0.041)			
MOA-funded	−0.005	0.058	0.261*	0.117	0.071
	(0.097)	(0.070)	(0.107)	(0.098)	(0.070)
FSS-provided	0.086	0.023	−0.207*	−0.181**	−0.186*
	(0.062)	(0.072)	(0.089)	(0.061)	(0.079)
Male	0.223*	−0.011	0.024	−0.037	0.204*
	(0.097)	(0.074)	(0.082)	(0.084)	(0.092)
Age	−0.008	0.001	0.008	−0.001	0.008
	(0.007)	(0.006)	(0.008)	(0.009)	(0.008)
Airman (E1–E3)	0.499***	−0.054	0.279*	−0.132	0.106
	(0.131)	(0.106)	(0.128)	(0.147)	(0.142)
Senior airman (E4)	0.089	−0.053	0.302	−0.003	0.206
	(0.199)	(0.094)	(0.186)	(0.181)	(0.170)
Senior noncommissioned officer (E7–E9)	0.277*	−0.106	0.220*	0.145	0.223*
	(0.114)	(0.096)	(0.111)	(0.120)	(0.102)
Company grade officer (O1–O3)	0.359**	0.102	0.270*	0.197	0.248
	(0.135)	(0.078)	(0.130)	(0.119)	(0.131)

	Social Cohesion First Post-UNITE Survey Direct Associations	Social Cohesion Second Post-UNITE Survey Direct Associations	Decompress Direct Associations	Social Interaction Direct Associations	Peer/Air Force Values Direct Associations
Field grade officer (O4–O6)	0.565***	0.160	0.250*	0.360**	0.276*
	(0.116)	(0.123)	(0.115)	(0.120)	(0.137)
Remote/isolated installation	0.229	0.191*	−0.108	0.018	−0.098
	(0.128)	(0.077)	(0.146)	(0.097)	(0.087)
Installation of medium size	0.168	0.058	0.121	0.096	−0.034
	(0.133)	(0.071)	(0.133)	(0.127)	(0.118)
Installation of large size	0.372*	0.095	0.115	0.137	0.060
	(0.159)	(0.081)	(0.171)	(0.182)	(0.146)
Installation of mega-large size	−0.036	0.007	0.119	0.023	0.101
	(0.190)	(0.147)	(0.185)	(0.169)	(0.129)
Cohesion, first post-event survey		0.632***			
		(0.041)			
Constant	−0.405	−0.214			
	(0.299)	(0.228)			
Observations	624	624	624	624	624

NOTE: Standard errors are clustered at the installation level. All building blocks and cohesion outcomes are standardized. Reference category includes noncommissioned officer (E7–E9), non-remote/isolated installation, and small installation size. * indicates $p < 0.05$; ** indicates $p < 0.01$; and *** indicates $p < 0.001$.

Table D.7. Indirect and Total Associations of Event Characteristics, Building Blocks, and Social Cohesion

	Social Cohesion First Post-UNITE Survey Indirect Associations	Social Cohesion Second Post-UNITE Survey Indirect Associations	Social Cohesion First Post-UNITE Survey Total Associations	Social Cohesion Second Post-UNITE Survey Total Associations
Decompress		0.065	0.103	0.178**
		(0.047)	(0.074)	(0.051)
Social interaction		0.090*	0.142**	0.006
		(0.035)	(0.054)	(0.062)
Peer/Air Force values		0.089**	0.140**	0.146**
		(0.033)	(0.052)	(0.047)
MOA-funded	0.053	0.054	0.048	0.112
	(0.031)	(0.069)	(0.103)	(0.118)
FSS-provided	−0.073	−0.011	0.013	0.012
	(0.026)	(0.050)	(0.072)	(0.094)

	Social Cohesion First Post-UNITE Survey Indirect Associations	Social Cohesion Second Post-UNITE Survey Indirect Associations	Social Cohesion First Post-UNITE Survey Total Associations	Social Cohesion Second Post-UNITE Survey Total Associations
Male	0.026	0.175**	0.249*	0.164
	(0.029)	(0.066)	(0.102)	(0.092)
Age	0.002	−0.002	−0.006	−0.001
	(0.003)	(0.005)	(0.008)	(0.008)
Airman (E1–E3)	0.025	0.379***	0.524***	0.326**
	(0.055)	(0.091)	(0.146)	(0.156)
Senior airman (E4)	0.060	0.140	0.149	0.087*
	(0.066)	(0.121)	(0.195)	(0.169)
Senior noncommissioned officer (E7–E9)	0.075*	0.247**	0.352**	0.142
	(0.034)	(0.079)	(0.124)	(0.126)
Company grade officer (O1–O3)	0.091*	0.312***	0.450**	0.414***
	(0.043)	(0.090)	(0.140)	(0.130)
Field grade officer (O4–O6)	0.116*	0.444***	0.680***	0.603
	(0.047)	(0.090)	(0.128)	(0.141)
Remote/isolated installation	−0.022	0.111	0.206	0.302*
	(0.036)	(0.091)	(0.129)	(0.125)
Installation of medium size	0.021	0.123	0.189	0.181
	(0.039)	(0.113)	(0.162)	(0.141)
Installation of large size	0.040	0.265*	0.411*	0.360**
	(0.057)	(0.124)	(0.184)	(0.133)
Installation of mega-large size	0.030	0.013	−0.006	0.020
	(0.053)	(0.142)	(0.211)	(0.151)
Cohesion, first post-event survey				0.632***
				(0.041)
Observations	624	624	624	624

NOTE: Standard errors are clustered at the installation level. All building blocks and cohesion outcomes are standardized. Reference category includes noncommissioned officer (E7–E9), non-remote/isolated installation, and small installation size. * indicates $p < 0.05$; ** indicates $p < 0.01$; and *** indicates $p < 0.001$.

Supplementary Analyses

Finally, all analyses in Chapter 4 and our findings regarding the main research questions use the sample of airmen who responded to both the first and second post-event surveys. For completeness, we present results from a larger sample of airmen that includes those who completed the first post-event survey but not the second post-event survey. For this sample, we could only analyze cohesion outcomes from the first post-event surveys, illustrated in Figure D.5.

Figure D.5. Supplemental Path Analysis Approach, Simple Model

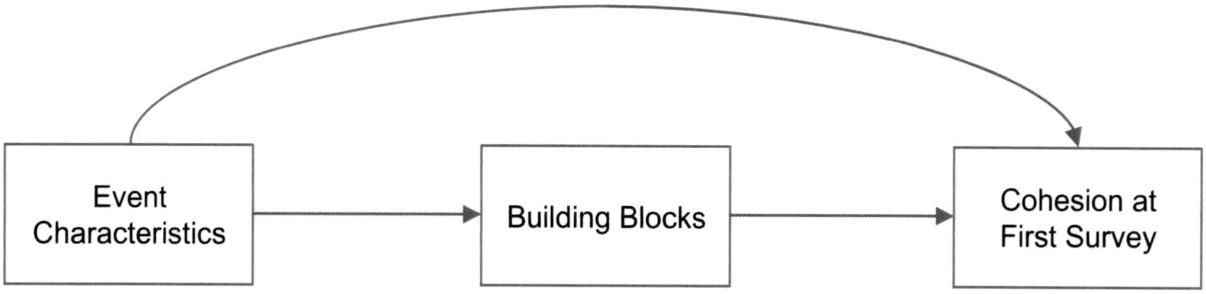

As in our previous models, our estimated model included all direct paths between event characteristics and building blocks, event characteristics and cohesion outcomes, and building blocks and cohesion outcomes. We also included the same vector of covariates and cluster standard errors at the installation level. We performed this analysis on the larger sample of airmen who took the first post-event survey (2,216 respondents) and the subsample who took both surveys (624) to show that estimates remain largely stable between the two samples.

Tables of Supplementary Analyses on Larger Sample of Airmen

In this section, we provide results on direct, indirect, and total associations between building event characteristics, building blocks, and overall cohesion for a larger sample that includes airmen who responded to the first, but not the second, post-event survey. The goal is to compare the stability of estimates across samples. Table D.8 presents the results for the larger sample of airmen (2,216 respondents), and Table D.9 presents the results for the sample used in the main analysis (624 respondents). In both samples, building blocks were positively and significant associated with cohesion, as seen in the main analyses. Similarly, FSS-provided events were negatively associated with building blocks and indirectly negatively associated with cohesion, as seen in the main models. The only difference is that, in the larger sample, MOA-funded events were also positively and significantly associated with building blocks and indirectly, positively, and significantly associated with cohesion. In the main sample, only the association between the decompression building block and MOA funding was significant.

Table D.8. Direct, Indirect, and Total Associations of Event Characteristics, Building Blocks, and Overall Cohesion Among Airmen Responding to First Post-Event Survey

	Cohesion First Post-Event Survey Direct Effects	Decompress Direct Effects	Social Interaction Direct Effects	Peer/Air Force Values Direct Effects	Cohesion First Post-Event Survey Indirect Effects	Cohesion First Post-Event Survey Total Effects
Decompress	0.122***					0.122***
	(0.031)					(0.031)
Social interaction	0.163***					0.163***
	(0.025)					(0.025)
Peer/Air Force values	0.202***					0.202***
	(0.025)					(0.025)
MOA-funded	0.034	0.366***	0.198***	0.170**	0.111***	0.145
	(0.068)	(0.063)	(0.056)	(0.054)	(0.025)	(0.077)
FSS-provided	0.072	−0.176**	−0.148***	−0.111*	−0.068**	0.004
	(0.045)	(0.062)	(0.039)	(0.053)	(0.023)	(0.052)
Male	0.234***	−0.052	−0.042	0.057	−0.002	0.233***
	(0.042)	(0.043)	(0.055)	(0.050)	(0.021)	(0.043)
Age	0.003	0.006	0.004	0.008	0.003	0.006
	(0.005)	(0.004)	(0.005)	(0.005)	(0.002)	(0.006)
Airman (E1–E3)	0.463***	0.162*	−0.021	0.024	0.021	0.484***
	(0.071)	(0.078)	(0.068)	(0.074)	(0.030)	(0.079)
Senior airman (E4)	0.069	0.110	−0.034	0.079	0.024	0.093
	(0.087)	(0.100)	(0.091)	(0.090)	(0.039)	(0.105)
Senior noncommissioned officer (E7–E9)	0.148*	0.267***	0.257***	0.175*	0.110**	0.258**
	(0.061)	(0.079)	(0.077)	(0.074)	(0.035)	(0.075)
Company grade officer (O1–O3)	0.396***	0.318***	0.351***	0.125	0.121***	0.518***
	(0.081)	(0.078)	(0.071)	(0.075)	(0.034)	(0.096)
Field grade officer (O4–O6)	0.392***	0.285***	0.412***	0.127	0.128**	0.520***
	(0.085)	(0.081)	(0.082)	(0.090)	(0.037)	(0.099)
Remote/isolated installation	0.159**	−0.030	0.016	−0.104*	−0.022	0.137*
	(0.055)	(0.078)	(0.047)	(0.053)	(0.021)	(0.064)
Installation of medium size	0.099	−0.019	−0.080	−0.131*	−0.042*	0.057
	(0.058)	(0.057)	(0.048)	(0.054)	(0.021)	(0.066)
Installation of large size	0.000	−0.050	−0.111	−0.140*	−0.052*	−0.052
	(0.060)	(0.054)	(0.061)	(0.062)	(0.026)	(0.064)

	Cohesion First Post-Event Survey Direct Effects	Decompress Direct Effects	Social Interaction Direct Effects	Peer/Air Force Values Direct Effects	Cohesion First Post-Event Survey Indirect Effects	Cohesion First Post-Event Survey Total Effects
Installation of mega-large size	0.006	−0.067	−0.174	−0.143	−0.066	−0.059
	(0.073)	(0.094)	(0.097)	(0.102)	(0.044)	(0.092)
Constant	−0.625***	−0.442*	−0.208	−0.291		
	(0.186)	(0.173)	(0.164)	(0.187)		
Observations	2,216	2,216	2,216	2,216	2,216	2,216

NOTE: Standard errors are clustered at the installation level. All building blocks and cohesion outcomes are standardized. Reference category includes noncommissioned officer (E7–E9), non-remote/isolated installation, and small installation size. * indicates $p < 0.05$; ** indicates $p < 0.01$; and *** indicates $p < 0.001$.

Table D.9. Direct, Indirect, and Total Associations of Event Characteristics, Building Blocks, and Overall Cohesion Among Airmen Responding to First and Second Post-Event Surveys

	Cohesion First Post-Event Survey Direct Effects	Decompress Direct Effects	Social Interaction Direct Effects	Peer Air Force Values Direct Effects	Cohesion First Post-Event Survey Indirect Effects	Cohesion First Post-Event Survey Total Effects
Decompress	0.184**					0.184**
	(0.058)					(0.058)
Social interaction	0.212***					0.212***
	(0.051)					(0.051)
Peer/Air Force values	0.165***					0.165***
	(0.045)					(0.045)
MOA-funded	−0.057	0.261*	0.117	0.071	0.084	0.027
	(0.100)	(0.107)	(0.098)	(0.070)	(0.045)	(0.113)
FSS-provided	0.056	−0.207*	−0.181**	−0.186*	−0.107**	−0.051
	(0.059)	(0.089)	(0.061)	(0.079)	(0.035)	(0.076)
Male	0.258**	0.024	−0.037	0.204*	0.030	0.288***
	(0.097)	(0.082)	(0.084)	(0.092)	(0.039)	(0.107)
Age	−0.002	0.008	−0.001	0.008	0.002	0.001
	(0.008)	(0.008)	(0.009)	(0.008)	(0.004)	(0.009)
Airman (E1–E3)	0.503***	0.279*	−0.132	0.106	0.041	0.544***
	(0.130)	(0.128)	(0.147)	(0.142)	(0.073)	(0.151)
Senior airman (E4)	0.083	0.302	−0.003	0.206	0.089	0.172
	(0.198)	(0.186)	(0.181)	(0.170)	(0.093)	(0.214)

	Cohesion First Post-Event Survey Direct Effects	Decompress Direct Effects	Social Interaction Direct Effects	Peer Air Force Values Direct Effects	Cohesion First Post-Event Survey Indirect Effects	Cohesion First Post-Event Survey Total Effects
Senior noncommissioned officer (E7–E9)	0.224*	0.220*	0.145	0.223*	0.108*	0.332**
	(0.099)	(0.111)	(0.120)	(0.102)	(0.052)	(0.120)
Company grade officer (O1–O3)	0.310*	0.270*	0.197	0.248	0.132*	0.442**
	(0.134)	(0.130)	(0.119)	(0.131)	(0.061)	(0.146)
Field grade officer (O4–O6)	0.519***	0.250*	0.360**	0.276*	0.168**	0.686***
	(0.115)	(0.115)	(0.120)	(0.137)	(0.064)	(0.139)
Remote/isolated installation	0.152	−0.108	0.018	−0.098	−0.032	0.120
	(0.110)	(0.146)	(0.097)	(0.087)	(0.053)	(0.121)
Installation of medium size	0.159	0.121	0.096	−0.034	0.037	0.196
	(0.128)	(0.133)	(0.127)	(0.118)	(0.057)	(0.163)
Installation of large size	0.286*	0.115	0.137	0.060	0.060	0.346*
	(0.143)	(0.171)	(0.182)	(0.146)	(0.083)	(0.163)
Installation of mega-large size	−0.083	0.119	0.023	0.101	0.043	−0.039
	(0.186)	(0.185)	(0.169)	(0.129)	(0.081)	(0.225)
Constant	−0.490	−0.581*	−0.035	−0.455		
	(0.317)	(0.284)	(0.291)	(0.291)		
Observations	624	624	624	624	624	624

NOTE: Standard errors are clustered at the installation level. All building blocks and cohesion outcomes are standardized. Reference category includes noncommissioned officer (E7–E9), non-remote/isolated installation, and small installation size. * indicates $p < 0.05$; ** indicates $p < 0.01$; and *** indicates $p < 0.001$.

References

Aberbach, Joel D., and Bert A. Rockman, "Conducting and Coding Elite Interviews," *PS: Political Science and Politics*, Vol. 35, No. 4, December 2002, pp. 673–676.

Acosta, Joie D., Rajeev Ramchand, Amariah Becker, Alexandria Felton, and Aaron Kofner, *RAND Suicide Prevention Program Evaluation Toolkit*, Santa Monica, Calif.: RAND Corporation, TL-111-OSD, 2013. As of October 27, 2017: https://www.rand.org/pubs/tools/TL111.html

Adler, Amy B., Jason Williams, Dennis McGurk, Andrew Moss, and Paul D. Bliese, "Resilience Training with Soldiers During Basic Combat Training: Randomisation by Platoon," *Applied Psychology: Health and Well-Being*; Vol. 7, No. 1, March 2015, pp. 85–107.

Aeron, Sapnaa, and Suman Pathak, "Personality Composition in Indian Software Teams and Its Relationship to Social Cohesion and Task Cohesion," *International Journal of Indian Culture and Business Management*, Vol. 13, No. 3, 2016, pp. 267–287.

Air Force Instruction 65-106, *Appropriated Fund Support of Morale, Welfare, and Recreation (MWR) and Other Nonappropriated Fund Instrumentalities (NAFIS)*, Washington, D.C.: Department of the Air Force, January 15, 2019.

Annessi, James J., "Effects of Minimal Group Promotion on Cohesion and Exercise Adherence," *Small Group Research*, Vol. 30, No. 5, October 1999, pp. 542–557.

Arthur, Calum Alexander, and Lew Hardy, "Transformational Leadership: A Quasi-Experimental Study," *Leadership and Organization Development Journal*, Vol. 35, No. 1, 2014, pp. 38–53.

Aubke, Florian, Karl Wöber, Noel Scott, and Rodolfo Baggio, "Knowledge Sharing in Revenue Management Teams: Antecedents and Consequences of Group Cohesion," *International Journal of Hospitality Management*, Vol. 41, August 2014, pp. 149–157.

Banning, Mary Rus, and David L. Nelson, "The Effects of Activity-Elicited Humor and Group Structure on Group Cohesion and Affective Responses," *American Journal of Occupational Therapy*, Vol. 41, No. 8, August 1987, pp. 510–514.

Barnett, Robert, "Revitalizing Squadrons, Air Force Outlines Progress," U.S. Air Force, August 10, 2018. As of May 22, 2020: https://www.af.mil/News/Article-Display/Article/1598301/revitalizing-squadrons-air-force-outlines-progress/

Barrett, Ann, Carolyn Piatek, Susan Korber, and Cynthia Padula, "Lessons Learned from a Lateral Violence and Team-Building Intervention," *Nursing Administration Quarterly*, Vol. 33, No. 4, October–December 2009, pp. 342–351.

Bartone, Paul T., Bjørn Helge Johnsen, Jarle Eid, Wibecke Brun, and Jon C. Laberg, "Factors Influencing Small-Unit Cohesion in Norwegian Navy Officer Cadets," *Military Psychology*, Vol. 14, No. 1, January 2002, pp. 1–22.

Beal, Daniel J., Robin R. Cohen, Michael J. Burke, and Christy L. McLendon, "Cohesion and Performance in Groups: A Meta-Analytic Clarification of Construct Relations," *Journal of Applied Psychology*, Vol. 88, No. 6, December 2003, pp. 989–1004.

Benson, Alex J., Mark A. Eys, and P. Gregory Irving, "Great Expectations: How Role Expectations and Role Experiences Relate to Perceptions of Group Cohesion," *Journal of Sport and Exercise Psychology*, Vol. 38, No. 2, April 2016, pp. 160–172.

Birx, Ellen, Kathleen B. Lasala, and Mark Wagstaff, "Evaluation of a Team-Building Retreat to Promote Nursing Faculty Cohesion and Job Satisfaction," *Journal of Professional Nursing*, Vol. 27, No. 3, May–June 2011, pp. 174–178.

Boer, Diana, and Amina Abubakar, "Music Listening in Families and Peer Groups: Benefits for Young People's Social Cohesion and Emotional Well-Being Across Four Cultures," *Frontiers in Psychology*, Vol. 5, 2014.

Boyd, Michael, Mi-Sook Kim, Nurcan Ensari, and Zenong Yin, "Perceived Motivational Team Climate in Relation to Task and Social Cohesion Among Male College Athletes," *Journal of Applied Social Psychology*, Vol. 44, No. 2, February 2014, pp. 115–123.

Brisimis, Evangelos, Evangelos Bebetsos, and Charalampos Krommidas, "Does Group Cohesion Predict Team Sport Athletes' Satisfaction?" *Hellenic Journal of Psychology*, Vol. 15, No. 1, 2018, pp. 108–124.

Brissett, Wilson, "Revitalizing USAF's Squadrons," *Air Force Magazine*, August 29, 2017. As of May 22, 2020:
https://www.airforcemag.com/article/Revitalizing-USAFs-Squadrons/

Bronfenbrenner, Urie, *The Ecology of Human Development: Experiments by Nature and Design*, Cambridge, Mass.: Harvard University Press, 1979.

Bruner, Mark William, and Kevin S. Spink, "Evaluating a Team Building Intervention in a Youth Exercise Setting," *Group Dynamics: Theory, Research, and Practice*, Vol. 14, No. 4, December 2010, pp. 304–317.

Bugen, Larry A., "Composition and Orientation Effects on Group Cohesion," *Psychological Reports*, Vol. 40, No. 1, 1977, pp. 175–181.

Caldwell, Linda L., "Leisure and Health: Why Is Leisure Therapeutic?" *British Journal of Guidance and Counselling*, Vol. 33, No. 1, 2005, pp. 7–26.

Carron, Albert V., and Lawrence R. Brawley, "Cohesion: Conceptual and Measurement Issues," *Small Group Research*, Vol. 31, No. 1, 2000, pp. 89–106.

Carron, Albert V., Lawrence R. Brawley, and W. Neil Widmeyer, "The Measurement of Cohesiveness in Sport Groups," in Joan L. Duda, ed., *Advances in Sport and Exercise Psychology Measurement*, Morgantown, W.V.: Fitness Information Technology, 1998, pp. 213–226.

Carron, Albert V., and Kevin S. Spink, "Team Building in an Exercise Setting," *Sport Psychologist*, Vol. 7, No. 1, March 1993, pp. 8–18.

Carron, Albert V., W. Neil Widmeyer, and Lawrence R. Brawley, "The Development of an Instrument to Assess Cohesion in Sport Teams: The Group Environment Questionnaire," *Journal of Sport and Exercise Psychology*, Vol. 7, No. 3, January 1985, pp. 244–266.

Chang, Artemis, and Prashant Bordia, "A Multidimensional Approach to the Group Cohesion-Group Performance Relationship," *Small Group Research*, Vol. 32, No. 4, August 2001, pp. 379–405.

Charbonneau, Danielle, and Valerie M. Wood, "Antecedents and Outcomes of Unit Cohesion and Affective Commitment to the Army," *Military Psychology*, Vol. 30, No. 1, February 2018, pp. 43–53.

Chen, Chun-Hsi Vivian, Ya-Yun Tang, and Shih-Jon Wang, "Interdependence and Organizational Citizenship Behavior: Exploring the Mediating Effect of Group Cohesion in Multilevel Analysis," *Journal of Psychology: Interdisciplinary and Applied*, Vol. 143, No. 6, December 2009, pp. 625–640.

Chen, Hsing-Jung, and Pamela J. Kovacs, "Working with Families in Which a Parent Has Depression: A Resilience Perspective," *Families in Society*, Vol. 94, No. 2, 2013, pp. 114–120.

Christensen, Ulla, Lone Schmidt, Esben Budtz-Jørgensen, and Kirsten Avlund, "Group Cohesion and Social Support in Exercise Classes: Results from a Danish Intervention Study," *Health Education and Behavior*, Vol. 33, No. 5, October 2006, pp. 677–689.

Cleirigh, Daire O., and John Greaney, "Mindfulness and Group Performance: An Exploratory Investigation into the Effects of Brief Mindfulness Intervention on Group Task Performance," *Mindfulness*, Vol. 6, No. 3, June 2015, pp. 601–609.

Clem, Jamie M., Thomas E. Smith, and Kristin V. Richards, "Effects of a Low-Element Challenge Course on Abstinence Self-Efficacy and Group Cohesion," *Research on Social Work Practice*, Vol. 22, No. 2, March 2012, pp. 151–158.

Cordobés, Tania K., "Group Songwriting as a Method for Developing Group Cohesion for HIV-Seropositive Adult Patients with Depression," *Journal of Music Therapy*, Vol. 34, No. 1, Spring 1997, pp. 46–67.

DeCuir-Gunby, Jessica T., Patricia L. Marshall, and Allison W. McCulloch, "Developing and Using a Codebook for the Analysis of Interview Data: An Example from a Professional Development Research Project," *Field Methods*, Vol. 23, No. 2, 2011, pp. 136–155.

Dermatis, Helen, Mary Salke, Marc Galanter, and Gregory Bunt, "The Role of Social Cohesion Among Residents in a Therapeutic Community," *Journal of Substance Abuse Treatment*, Vol. 21, No. 2, September 2001, pp. 105–110.

DiCicco-Bloom, Barbara, and Benjamin F. Crabtree, "The Qualitative Research Interview," *Medical Education*, Vol. 40, No. 4, April 2006, pp. 314–321.

DiMeglio, Karen, Cynthia Padula, Carolyn Piatek, Susan Korber, Ann Barrett, Maria Ducharme, Sandra Lucas, Nicole Piermont, Elaine Joyal, Virginia DeNicola, and Karen Corry, "Group Cohesion and Nurse Satisfaction: Examination of a Team-Building Approach," *Journal of Nursing Administration*, Vol. 35, No. 3, March 2005, pp. 110–120.

Durdubas, Deniz, Luc J. Martin, and Ziya Koruc, "A Season-Long Goal-Setting Intervention for Elite Youth Basketball Teams," *Journal of Applied Sport Psychology*, Vol. 32, No. 6, 2020, pp. 529–545.

Eatough, Erin, Chu-Hsiang Chang, and Nicholas Hall, "Getting Roped In: Group Cohesion, Trust, and Efficacy Following a Ropes Course Intervention," *Performance Improvement Quarterly*, Vol. 28, No. 2, July 2015, pp. 65–89.

Evans, Charles R., and Kenneth L. Dion, "Group Cohesion and Performance: A Meta-Analysis," *Small Group Research*, Vol. 43, No. 6, December 2012, pp. 690–701.

Forrest, Christopher K., and Mark W. Bruner, "Evaluating Social Media as a Platform for Delivering a Team-Building Exercise Intervention: A Pilot Study," *International Journal of Sport and Exercise Psychology*, Vol. 15, No. 2, 2017, pp. 190–206.

Galyon, Charles E., Eleanore C. T. Heaton, Tiffany L. Best, and Robert L. Williams, "Comparison of Group Cohesion, Class Participation, and Exam Performance in Live and Online Classes," *Social Psychology of Education*, Vol. 19, No. 1, March 2016, pp. 61–76.

García-Guiu, Carlos, Fernando Molero, and Juan A. Moriano, "Authentic Leadership and its Influence on Group Cohesion and Organizational Identification: The Role of Organizational Justice as a Mediating Variable," *Revista de Psicología Social*, Vol. 30, No. 1, January 2015, pp. 60–88.

Goldfein, David L., "CSAF Letter to Airmen," U.S. Air Force, August 9, 2016. As of May 22, 2020:
https://www.af.mil/News/Article-Display/Article/873161/csaf-letter-to-airmen/

Göritz, Anja S., and Miriam Rennung, "Interpersonal Synchrony Increases Social Cohesion, Reduces Work-Related Stress and Prevents Sickdays: A Longitudinal Field Experiment," *Gruppe. Interaktion. Organisation.*, Vol. 50, No. 1, March 2019, pp. 83–94.

Grady, Joe, Alyssa Banford-Witting, Angela Kim, and Sean Davis, "Differences in Unit Cohesion and Combat-Related Mental Health Problems Based on Attachment Styles in US Military Veterans," *Contemporary Family Therapy*, Vol. 40, No. 3, September 2018, pp. 249–258.

Griffith, James, "The Army's New Unit Personnel Replacement and Its Relationship to Unit Cohesion and Social Support," *Military Psychology*, Vol. 1, No. 1, 1989, pp. 17–34.

Gupta, Vishal K., Rui Huang, and Suman Niranjan, "A Longitudinal Examination of the Relationship Between Team Leadership and Performance," *Journal of Leadership and Organizational Studies*, Vol. 17, No. 4, November 2010, pp. 335–350.

Harrison, David A., Kenneth H. Price, and Myrtle P. Bell, "Beyond Relational Demography: Time and the Effects of Surface- and Deep-Level Diversity on Work Group Cohesion," *Academy of Management Journal*, Vol. 41, No. 1, February 1998, pp. 96–107.

Heuzé, Jean-Philippe, Grégoire Bosselut, and Jean-Philippe Thomas, "Should the Coaches of Elite Female Handball Teams Focus on Collective Efficacy or Group Cohesion?" *Sport Psychologist*, Vol. 21, No. 4, December 2007, pp. 383–399.

Holland, Jessica N., and Adam T. Schmidt, "Static and Dynamic Factors Promoting Resilience Following Traumatic Brain Injury: A Brief Review," *Neural Plasticity*, Vol. 2015, 2015.

Hoogstraten, Johan, and Harrie C. M. Vorst, "Group Cohesion, Task Performance, and the Experimenter Expectancy Effect," *Human Relations*, Vol. 31, No. 11, November 1978, pp. 939–956.

Hopkins-Chadwick, Denise L., "The Health Readiness of Junior Enlisted Military Women: The Social Determinants of Health Model and Research Questions," *Military Medicine*, Vol. 171, No. 6, June 2006, pp. 544–549.

Hughes, Richard L., William E. Rosenbach, and William H. Clover, "Team Development in an Intact, Ongoing Work Group: A Quasi-Field Experiment," *Group and Organization Studies*, Vol. 8, No. 2, June 1983, pp. 161–186.

Ibbetson, Adrian, and Sue Newell, "A Comparison of a Competitive and Non-Competitive Outdoor Management Development Programme," *Personnel Review*, Vol. 28, No. 1/2, 1999, pp. 58–76.

Irwin, Brandon, Daniel Kurz, Patrice Chalin, and Nicholas Thompson, "Testing the Efficacy of OurSpace, a Brief, Group Dynamics-Based Physical Activity Intervention: A Randomized Controlled Trial," *Journal of Medical Internet Research*, Vol. 18, No. 5, May 2016.

Ismail, Maimunah, Nordahlia Umar Baki, and Zoharah Omar, "The Influence of Organizational Culture and Organizational Justice on Group Cohesion as Perceived by Merger and Acquisition Employees," *Organizations and Markets in Emerging Economies*, Vol. 9, No. 2, December 2018, pp. 233–250.

Iwasaki, Yoshitaka, Jennifer MacTavish, and Kelly MacKay, "Building on Strengths and Resilience: Leisure as a Stress Survival Strategy," *British Journal of Guidance and Counseling*, Vol. 33, No. 1, February 2005, pp. 81–100.

Iwasaki, Yoshitaka, Roger C. Mannell, Bryan J. A. Smale, and Janice Butcher, "A Short-Term Longitudinal Analysis of Leisure Coping Used by Police and Emergency Response Service Workers," *Journal of Leisure Research*, Vol. 34, No. 3, September 2002, pp. 311–339.

Jenkins, Jayne M., and Brandon L. Alderman, "Influence of Sport Education on Group Cohesion in University Physical Education," *Journal of Teaching in Physical Education*, Vol. 30, No. 3, July 2011, pp. 214–230.

Jowett, Sophia, and Victoria Chaundy, "An Investigation into the Impact of Coach Leadership and Coach-Athlete Relationship on Group Cohesion," *Group Dynamics: Theory, Research, and Practice*, Vol. 8, No. 4, December 2004, pp. 302–311.

Kawachi, Ichiro, and Lisa F. Berkman, "Social Cohesion, Social Capital, and Health," in Lisa F. Berkman and Ichiro Kawachi, eds., *Social Epidemiology*, New York: Oxford University Press, 2000, pp. 174–190.

Kaymak, Tuhan, "Group Cohesion and Performance: A Search for Antecedents," *E a M: Ekonomie a Management*, Vol. 14, No. 4, 2011, pp. 78–91.

Kearns, Ade, and Ray Forrest, "Social Cohesion and Multilevel Urban Governance," *Urban Studies*, Vol. 37, No. 5–6, 2000, pp. 995–1017.

Kim, Son Chae, Jaynelle F. Stichler, Laurie Ecoff, Caroline E. Brown, Ana-Maria Gallo, and Judy E. Davidson, "Predictors of Evidence-Based Practice Implementation, Job Satisfaction, and Group Cohesion Among Regional Fellowship Program Participants," *Worldviews on Evidence-Based Nursing*, Vol. 13, No. 5, October 2016, pp. 340–348.

Ko, Yu Kyung, "Group Cohesion and Social Support of the Nurses in a Special Unit and a General Unit in Korea," *Journal of Nursing Management*, Vol. 19, No. 5, July 2011, pp. 601–610.

Lafferty, M. E., C. Wakefield, and H. Brown, "'We Do It for the Team' – Student-Athletes' Initiation Practices and Their Impact on Group Cohesion," *International Journal of Sport and Exercise Psychology*, Vol. 15, No. 4, 2017, pp. 438–446.

Lee, Cynthia, and Jiing-Lih Farh, "Joint Effects of Group Efficacy and Gender Diversity on Group Cohesion and Performance," *Applied Psychology*, Vol. 53, No. 1, January 2004, pp. 136–154.

Levitt, Heidi M., Michael Bamberg, John W. Creswell, David M. Frost, Ruthellen Josselson, and Carola Suárez-Orozco, "Journal Article Reporting Standards for Qualitative Primary, Qualitative Meta-Analytic, and Mixed Methods Research in Psychology: The APA Publications and Communications Board Task Force Report," *American Psychologist*, Vol. 73, No. 1, 2018, pp. 26–46.

Lin, Cheng-Chen, and Tai-Kuang Peng, "From Organizational Citizenship Behaviour to Team Performance: The Mediation of Group Cohesion and Collective Efficacy," *Management and Organization Review*, Vol. 6, No. 1, March 2010, pp. 55–75.

López, Carlos García-Guiu, Fernando Molero Alonso, Miguel Moya Morales, and Juan Antonio Moriano León, "Authentic Leadership, Group Cohesion and Group Identification in Security and Emergency Teams," *Psicothema*, Vol. 27, No. 1, 2015, pp. 59–64.

Maxwell, Joseph A., "Using Numbers in Qualitative Research," *Qualitative Inquiry*, Vol. 16, No. 6, July 2010, pp. 475–482.

McFadden, Paula, Anne Campbell, and Brian Taylor, "Resilience and Burnout in Child Protection Social Work: Individual and Organisational Themes from a Systematic Literature Review," *British Journal of Social Work*, Vol. 45, No. 5, July 2015, pp. 1546–1563.

McGonigle, Timothy P., Wendy J. Casper, Edward P. Meiman, Candace Blair Cronin, Brian E. Cronin, and Rebecca R. Harris, "The Relationship Between Personnel Support Programs and Readiness: A Model to Guide Future Research," *Military Psychology*, Vol. 17, No. 1, 2005, pp. 25–39.

McLaren, Colin D., Mark A. Eys, and Robyn A. Murray, "A Coach-Initiated Motivational Climate Intervention and Athletes' Perceptions of Group Cohesion in Youth Sport," *Sport, Exercise, and Performance Psychology*, Vol. 4, No. 2, 2015, pp. 113–126.

Meadows, Sarah O., Stephanie Brooks Holliday, Wing Yi Chan, Stephani L. Wrabel, Margaret Tankard, Dana Schultz, Christopher M. Busque, Felix Knutson, Leslie Adrienne Payne, and Laura L. Miller, *Air Force Morale, Welfare, and Recreation Programs and Services: Contribution to Airman and Family Resilience and Readiness*, Santa Monica, Calif.: RAND Corporation, RR-2670-AF, 2019. As of January 19, 2021: https://www.rand.org/pubs/research_reports/RR2670.html

Meyer, Barbara B., "The Ropes and Challenge Course: A Quasi-Experimental Examination," *Perceptual and Motor Skills*, Vol. 90, No. 3, June 2000, pp. 1249–1257.

Miles, Matthew B., and A. Michael Huberman, *Qualitative Data Analysis: An Expanded Sourcebook*, 2nd ed., Thousand Oaks, Calif.: Sage Publications, 1994.

Milstein, Bobby, Scott Wetterhall, and CDC Evaluation Working Group, "A Framework Featuring Steps and Standards for Program Evaluation," *Health Promotion Practice*, Vol. 1, No. 3, July 2000, pp. 221–228.

Moore, Amanda, and Ketevan Mamiseishvili, "Examining the Relationship Between Emotional Intelligence and Group Cohesion," *Journal of Education for Business*, Vol. 87, No. 5, 2012, pp. 296–302.

Morgan, Paul B. C., David Fletcher, and Mustafa Sarkar, "Recent Developments in Team Resilience Research in Elite Sport," *Current Opinion in Psychology*, Vol. 16, August 2017, pp. 159–164.

National Defense Research Institute, *Sexual Orientation and U.S. Military Personnel Policy: An Update of RAND's 1993 Study*, Santa Monica, Calif: RAND Corporation, MG-1056-OSD, 2010. As of January 19, 2021:
https://www.rand.org/pubs/monographs/MG1056.html

Neale, Joanna, Peter Miller, and Robert West, "Reporting Quantitative Information in Qualitative Research: Guidance for Authors and Reviewers," *Addiction*, Vol. 109, No. 2, February 2014, pp. 175–176.

Nilsson, Marco, "Primary Unit Cohesion Among the Peshmerga and Hezbollah," *Armed Forces and Society*, Vol. 44, No. 4, 2018, pp. 647–665.

Oh, Seungbin, Michelle D. Mitchell, Caitlyn McKinzie Bennett, Laura Rendon Finnell, Yvette Saliba, Nevin J. Heard, and Elizabeth R. Pennock, "Journal Sharing on Group Cohesion and Goal Attainment in Experiential Growth Groups," *Journal for Specialists in Group Work*, Vol. 43, No. 3, July 2018, pp. 206–229.

Oliver, Laurel W., Joan Harman, Elizabeth Hoover, Stephanie M. Hayes, and Nancy A. Pandhi, "A Quantitative Integration of the Military Cohesion Literature," *Military Psychology*, Vol. 11, No. 1, 1999, pp. 57–83.

Pack, D. Glenn, and Henry C. Rickard, "Self-Reports of Group Cohesion Under High and Low Cooperation," *Psychological Reports*, Vol. 36, No. 1, February 1975, p. 86.

Plante, Pierre, "Promoting Group Cohesion Through Art Therapy: A Project Adapted for Adults in a Community Centre," *Canadian Art Therapy Association Journal*, Vol. 19, No. 2, 2006, pp. 2–11.

Prapavessis, Harry, Albert V. Carron, and Kevin S. Spink, "Team Building in Sport," *International Journal of Sport Psychology*, Vol. 27, No. 3, July 1996, pp. 269–285.

Rohe, Daniel E., Patricia A. Barrier, Matthew M. Clark, David A. Cook, Kristin S. Vickers, and Paul A. Decker, "The Benefits of Pass-Fail Grading on Stress, Mood, and Group Cohesion in Medical Students," *Mayo Clinic Proceedings*, Vol. 81, No. 11, November 2006, pp. 1443–1448.

Rosen, Leora N., Paul D. Bliese, Kathleen A. Wright, and Robert K. Gifford, "Gender Composition and Group Cohesion in U.S. Army Units: A Comparison Across Five Studies," *Armed Forces and Society*, Vol. 25, No. 3, Spring 1999, pp. 365–386.

Rostker, Bernard D., Scott A. Harris, James P. Kahan, Erik J. Frinking, C. Neil Fulcher, Lawrence M. Hanser, Paul Koegel, John D. Winkler, Brent A. Boultinghouse, Joanna Heilbrunn, Janet Lever, Robert J. MacCoun, Peter Tiemeyer, Gail L. Zellman, Sandra H. Berry, Jennifer Hawes-Dawson, Samantha Ravich, Steven L. Schlossman, Timothy Haggarty, Tanjam Jacobson, Ancella Livers, Sherie Mershon, Andrew Cornell, Mark A. Schuster, David E. Kanouse, Raynard Kington, Mark Litwin, Conrad Peter Schmidt, Carl H. Builder, Peter Jacobson, Stephen A. Saltzburg, Roger Allen Brown, William Fedorochko, Marilyn Fisher Freemon, John F. Peterson, and James A. Dewar, *Sexual Orientation and U.S. Military Personnel Policy: Options and Assessment*, Santa Monica, Calif.: RAND Corporation, MR-323-OSD, 1993. As of August 6, 2019: https://www.rand.org/pubs/monograph_reports/MR323.html

Saldaña, Johnny, *The Coding Manual for Qualitative Researchers*, 3rd ed., Thousand Oaks, Calif.: Sage Publications, 2015.

Sánchez, Jose C., and Amaia Yurrebaso, "Group Cohesion: Relationships with Work Team Culture," *Psicothema*, Vol. 21, No. 1, March 2009, pp. 97–104.

Sanderson, Brooke, and Margo Brewer, "What Do We Know About Student Resilience in Health Professional Education? A Scoping Review of the Literature," *Nurse Education Today*, Vol. 58, November 2017, pp. 65–71.

Senécal, Julie, Todd M. Loughead, and Gordon A. Bloom, "A Season-Long Team-Building Intervention: Examining the Effect of Team Goal Setting on Cohesion," *Journal of Sport and Exercise Psychology*, Vol. 30, No. 2, April 2008, pp. 186–199.

Shields, David Lyle Light, Douglas E. Gardner, Brenda Jo Light Bredemeier, and Alan Bostro, "The Relationship Between Leadership Behaviors and Group Cohesion in Team Sports," *Journal of Psychology: Interdisciplinary and Applied*, Vol. 131, No. 2, 1997, pp. 196–210.

Spink, Kevin S., and Albert V. Carron, "The Effects of Team Building on the Adherence Patterns of Female Exercise Participants," *Journal of Sport and Exercise Psychology*, Vol. 15, No. 1, 1993, pp. 39–49.

Steen, Sam, Elaina Vasserman-Stokes, and Rachel Vannatta, "Group Cohesion in Experiential Growth Groups," *Journal for Specialists in Group Work*, Vol. 39, No. 3, 2014, pp. 236–256.

Stokes, Joseph Powell, "Components of Group Cohesion: Intermember Attraction, Instrumental Value, and Risk Taking," *Small Group Research*, Vol. 14, No. 2, May 1983, pp. 163–173.

Tiryaki, M. Şefik, and Fatma Çepikkurt, "Relations of Attachment Styles and Group Cohesion in Premier League Female Volleyball Players," *Perceptual and Motor Skills*, Vol. 104, No. 1, February 2007, pp. 69–78.

Treadwell, Thomas W., Emily E. Reisch, Letitia E. Travaglini, and V. K. Kumar, "The Effectiveness of Collaborative Story Building and Telling in Facilitating Group Cohesion in a College Classroom Setting," *International Journal of Group Psychotherapy*, Vol. 61, No. 4, October 2011, pp. 503–517.

Troth, Ashlea C., Peter J. Jordan, and Sandra A. Lawrence, "Emotional Intelligence, Communication Competence, and Student Perceptions of Team Social Cohesion," *Journal of Psychoeducational Assessment*, Vol. 30, No. 4, August 2012, pp. 414–424.

Turk, Herman, "Social Cohesion Through Variant Values: Evidence from Medical Role Relations," *American Sociological Review*, Vol. 28, No. 1, February 1963, pp. 28–37.

U.S. Air Force, "Air Force Services Agency," fact sheet, December 1, 2002. As of June 15, 2020: https://www.af.mil/About-Us/Fact-Sheets/Display/Article/104541/air-force-services-agency/

U.S. Department of Defense, *2018 Demographics: Profile of the Military Community*, Washington, D.C., 2019.

Vaitkus, Mark, and James Griffith, "An Evaluation of Unit Replacement on Unit Cohesion and Individual Morale in the U.S. Army All-Volunteer Force," *Military Psychology*, Vol. 2, No. 4, 1990, pp. 221–239.

van Vianen, Annelies E. M., and Carsten K. W. De Dreu, "Personality in Teams: Its Relationship to Social Cohesion, Task Cohesion, and Team Performance," *European Journal of Work and Organizational Psychology*, Vol. 10, No. 2, June 2001, pp. 97–120.

Watson, Jocelyn D., Kathleen A. Martin Ginis, and Kevin S. Spink, "Team Building in an Exercise Class for the Elderly," *Activities, Adaptation, and Aging*, Vol. 28, No. 3, September 2004, pp. 35–47.

Webber, Sheila Simsarian, and Lisa M. Donahue, "Impact of Highly and Less Job-Related Diversity on Work Group Cohesion and Performance: A Meta-Analysis," *Journal of Management*, Vol. 27, No. 2, April 2001, pp. 141–162.

Welsh, Janet A., Jonathan R. Olson, and Daniel F. Perkins, "Gender Differences in Post-Deployment Adjustment of Air Force Personnel: The Role of Wartime Experiences, Unit Cohesion, and Self-Efficacy," *Military Medicine*, Vol. 184, No. 1–2, January–February 2019, pp. e229–e234.

Widmeyer, W. Neil, Lawrence R. Brawley, and Albert V. Carron, *The Measurement of Cohesion in Sport Teams: The Group Environment Questionnaire*, London, Ontario: Sports Dynamics, 1985.

Wilson, Stuart, Evangelia Bassiou, Aysel Denli, Lynsey C. Dolan, and Matthew Watson, "Traveling Groups Stick Together: How Collective Directional Movement Influences Social Cohesion," *Evolutionary Psychology*, Vol. 16, No. 3, July 2018.

Windsor, Phyllis M., Jamie Barker, and Paul McCarthy, "Doing Sport Psychology: Personal-Disclosure Mutual-Sharing in Professional Soccer," *Sport Psychologist*, Vol. 25, April 2011, pp. 94–114.

Yousafzai, Aisha K., Muneera A. Rasheed, and Zulfiqar A. Bhutta, "Annual Research Review: Improved Nutrition—A Pathway to Resilience," *Journal of Child Psychology and Psychiatry*, Vol. 54, No. 4, April 2013, pp. 367–377.

Zolkoski, Staci M., and Lyndal M. Bullock, "Resilience in Children and Youth: A Review," *Children and Youth Services Review*, Vol. 34, No. 12, December 2012, pp. 2295–2303.